REVISE EDEXCEL GCSE (9–1)
Mathematics
Higher

REVISION WORKBOOK

Series Consultant: Harry Smith

Author: Navtej Marwaha

Also available to support your revision:

Revise GCSE Study Skills Guide   9781447967071

The **Revise GCSE Study Skills Guide** is full of tried-and-trusted hints and tips for how to learn more effectively. It gives you techniques to help you achieve your best – throughout your GCSE studies and beyond!

Revise GCSE Revision Planner   9781447967828

The **Revise GCSE Revision Planner** helps you to plan and organise your time, step-by-step, throughout your GCSE revision. Use this book and wall chart to mastermind your revision.

> **For the full range of Pearson revision titles across KS2, KS3, GCSE, AS/A Level and BTEC visit:**
> www.pearsonschools.co.uk/revise

Contents

A small bit of small print

Edexcel publishes Sample Assessment Material and the Specification on its website. This is the official content and this book should be used in conjunction with it. The questions in 'Now try this' have been written to help you practise every topic in the book. Remember: the real exam questions may not look like this.

Factors and primes

1 (a) Write the following numbers as products of powers of their prime factors.

> Circle any prime numbers – that's the end of a branch.

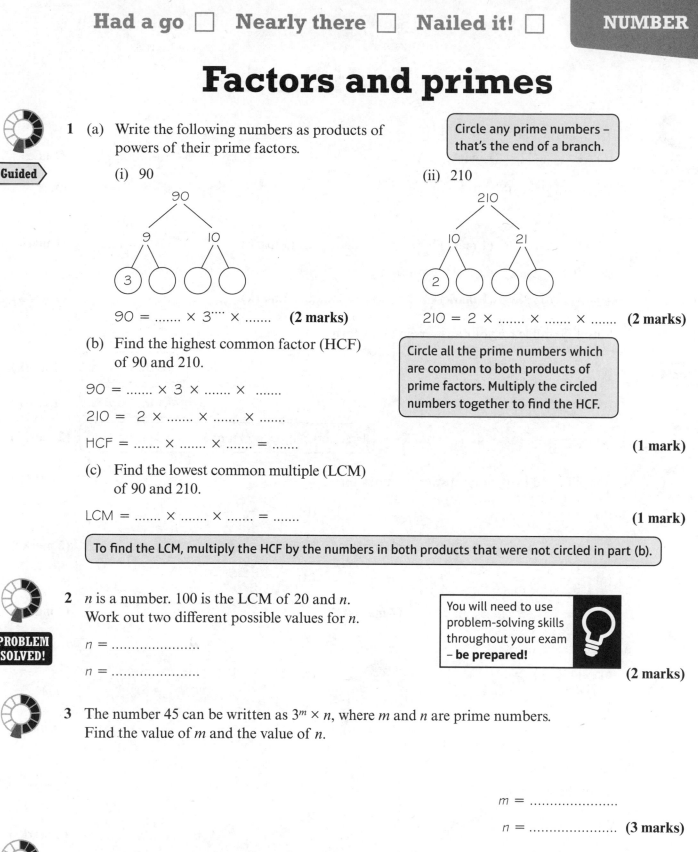

(i) 90

90 = ... × 3·⋯ × .... **(2 marks)**

(ii) 210

210 = 2 × .... × .... × .... **(2 marks)**

(b) Find the highest common factor (HCF) of 90 and 210.

90 = ... × 3 × .... × ....

210 = 2 × .... × .... × ....

HCF = ... × .... × .... = ... **(1 mark)**

> Circle all the prime numbers which are common to both products of prime factors. Multiply the circled numbers together to find the HCF.

(c) Find the lowest common multiple (LCM) of 90 and 210.

LCM = ... × .... × .... = ... **(1 mark)**

> To find the LCM, multiply the HCF by the numbers in both products that were not circled in part (b).

2 n is a number. 100 is the LCM of 20 and n.
Work out two different possible values for n.

> You will need to use problem-solving skills throughout your exam – **be prepared!**

n = ...

n = ... **(2 marks)**

3 The number 45 can be written as $3^m × n$, where m and n are prime numbers.
Find the value of m and the value of n.

m = ...

n = ... **(3 marks)**

4 A ramblers club buys hats in packs of 12 and scarves in packs of 18. The club buys exactly the same number of hats and scarves.
What is the smallest number of packs of hats and the smallest number of packs of scarves the club buys?

.... **(3 marks)**

Indices 1

Guided

1 Write as a single power of 4

(a) $4 \times 4 = 4^{\cdots}$ **(1 mark)**

(b) $4 \times 4 \times 4 \times 4 \times 4 = $ .... **(1 mark)**

2 Work out the value of

(a) 4^2

.... **(1 mark)**

(b) 2^3

.... **(1 mark)**

(c) $\sqrt{64}$

.... **(1 mark)**

(d) $\sqrt[3]{64}$

.... **(1 mark)**

(e) $\sqrt[3]{27}$

.... **(1 mark)**

(f) $\sqrt[3]{-64}$

.... **(1 mark)**

3 Simplify and leave your answers in index form.

Guided

(a) $5^3 \times 5^6 = 5^{3+6} = 5^{\cdots}$ | Add the powers. | **(1 mark)**

(b) $5^9 \div 5^6 = 5^{9-6} = 5^{\cdots}$ | Subtract the powers. | **(1 mark)**

(c) $\dfrac{5^{12}}{5 \times 5^7} = $ .... | First work out the power of 5 in the denominator. | **(2 marks)**

4 Simplify and leave your answers in index form.

(a) $\dfrac{3^2 \times 3^6}{3^5}$

.... **(2 marks)**

(b) $\dfrac{3^{12}}{3^6 \times 3^4}$

.... **(2 marks)**

(c) $\dfrac{3^7 \times 3^6}{3 \times 3^4}$

.... **(2 marks)**

(d) $\dfrac{3^8 \times 3^{-6}}{3 \times 3^{-5}}$

.... **(2 marks)**

5 Find the value of x.

Guided

(a) $11^3 \times 11^x = 11^{12}$

$11^{3+x} = 11^{12}$

$x = $ .... **(1 mark)**

(b) $11^{12} \div 11^x = 11^8$

$x = $ .... **(1 mark)**

6 $7^4 \times 7^x = \dfrac{7^9 \times 7^6}{7^3}$

Find the value of x.

$x = $ .... **(2 marks)**

7 Tom carried out an investigation and concluded that, '6 is a cube number since $2^3 = 6$.'

Is he correct? Explain your answer.

| You can explain your answer by writing a sentence with your reason, or by showing some neat working. |

Guided

No, because $2 \times 2 \times 2 = $ .... **(2 marks)**

8 $3^x \times 3^y = 3^{12}$ and $3^x \div 3^y = 3^2$

Work out the value of x and the value of y.

$x = $ .... $y = $ .... **(3 marks)**

Indices 2

1 Work out the value of

(a) 2^{-3} (b) 3^{-1} (c) 7^{-2} (d) $4^{-\frac{1}{2}}$

$\dfrac{1}{2^3} = \dfrac{\cdots}{\cdots}$ **(1 mark)** .... **(1 mark)** $\dfrac{1}{\cdots^2} = \dfrac{\cdots}{\cdots}$ **(1 mark)** .... **(1 mark)**

2 Work out the reciprocal of

(a) 3 (b) $\frac{1}{4}$ (c) $\frac{3}{5}$ (d) $\frac{9}{7}$

.... **(1 mark)** .... **(1 mark)** .... **(1 mark)** .... **(1 mark)**

3 Work out the value of

(a) $\left(\frac{2}{3}\right)^2$ (b) $\left(\frac{4}{3}\right)^3$ (c) $\left(\frac{4}{5}\right)^2$ (d) $\left(\frac{1}{5}\right)^3$

.... **(1 mark)** .... **(1 mark)** .... **(1 mark)** .... **(1 mark)**

4 Work out the value of

> Turn the fraction upside down, then change the negative power to a positive power.

(a) $\left(\dfrac{4}{3}\right)^{-2} = \left(\dfrac{3}{4}\right)^2 = \dfrac{3^2}{4^2} = \dfrac{\cdots}{\cdots}$ **(1 mark)** (b) $\left(\dfrac{1}{3}\right)^{-3} = \left(\dfrac{\cdots}{\cdots}\right)^3 = \dfrac{\cdots^3}{\cdots^3} = \cdots$ **(1 mark)**

(c) $\left(\dfrac{6}{5}\right)^{-2}$ (d) $\left(\dfrac{3}{5}\right)^{-3}$

.... **(1 mark)** .... **(1 mark)**

5 Work out the value of

(a) $25^{\frac{1}{2}}$ (b) $8^{\frac{1}{3}}$ (c) $64^{\frac{1}{3}}$ (d) $81^{\frac{1}{4}}$

.... **(1 mark)** .... **(1 mark)** .... **(1 mark)** .... **(1 mark)**

6 Work out the value of

(a) $16^{\frac{3}{2}}$ (b) $16^{\frac{3}{4}}$

$(16^{\frac{1}{2}})^3 = (\cdots)^3 = \cdots$ **(1 mark)** $(16^{\frac{1}{\cdots}})^{\cdots} = (\cdots)^{\cdots} = \cdots$ **(1 mark)**

(c) $25^{\frac{3}{2}}$ (d) $27^{\frac{2}{3}}$

$(25^{\frac{\cdots}{\cdots}})^{\cdots} = (\cdots)^{\cdots} = \cdots$ **(1 mark)** .... **(1 mark)**

7 Show that $8^{\frac{2}{3}} = 16^{\frac{1}{2}}$

> You will need to use problem-solving skills throughout your exam – **be prepared!**

(2 marks)

PROBLEM SOLVED!

8 $x = 3^m$ and $y = 3^n$

Express in terms of x and y

(a) 3^{m+n} (b) 3^{2n}

.... **(1 mark)** .... **(1 mark)**

Calculator skills 1

Guided

1 Work out, in each case giving your answer correct to 3 significant figures

> Use BIDMAS to remember the correct order of operations:
> **B**rackets
> **I**ndices
> **D**ivision
> **M**ultiplication
> **A**ddition
> **S**ubtraction

(a) $(11 + 8 \div 2)^3$

$(11 + ...)^3 = ...$ **(1 mark)**

(b) $(2 + 9 \times 10 + 3)^{\frac{1}{2}}$

.... **(1 mark)**

(c) $(8 + (3 \times 20) \div 6)^{\frac{2}{3}}$

.... **(1 mark)**

2 Work out

(a) $\dfrac{(27 + 3 \times 3)^2}{3 \times 2}$

(b) $\dfrac{(13 - \sqrt{12} \div 4)^3}{(4 + 3 \times 2)^2}$

.... **(1 mark)**   .... **(1 mark)**

3 Find the value of $\dfrac{4.5 + 3.75}{3.2^2 - 5.53}$

Write down all the figures on your calculator display.

Guided

$\dfrac{8.25}{..............} = ...$ **(2 marks)**

4 (a) Find the value of $\sqrt{30.25} + 1.75^2$

> Use your calculator to work out $\sqrt{30.25}$ and 1.75^2 separately. Write your answers before adding them. You might need to press the $\boxed{\text{S} \Leftrightarrow \text{D}}$ button to get your answer as a decimal number.

.... **(2 marks)**

(b) Write your answer to part (a) correct to 1 significant figure.

.... **(1 mark)**

5 $m = 7.1 \times 10^6$ and $n = 3.2 \times 10^{-3}$

Work out, in each case giving your answer in standard form correct to 3 significant figures

(a) mn

(b) $\dfrac{m}{n}$

.... **(2 marks)**   .... **(2 marks)**

6 Work out, in each case giving your answer correct to 3 significant figures

(a) $\sqrt{5.3} + \tan 38°$

(b) $\dfrac{288.3 \times \cos 58°}{(4.23 - 1.13)^3}$

.... **(2 marks)**   .... **(2 marks)**

7 $t^3 = \dfrac{mn}{m - n}$ $m = 4 \times 10^{12}$ $n = 3 \times 10^9$

Work out t. Give your answer in standard form correct to 3 significant figures.

$t = ...$ **(3 marks)**

Fractions

1 Work out

> Add the whole numbers first.

(a) $3\frac{4}{5} + 2\frac{3}{4} = 3 + 2 + \frac{4}{5} + \frac{3}{4}$

$= \text{.........} + \frac{\text{.....}}{20} + \frac{\text{.....}}{20} = \text{.........} + \frac{\text{.....}}{20} - \text{.........} + \text{.........}\frac{\text{.....}}{20} = \text{.........}\frac{\text{.....}}{20}$ **(3 marks)**

(b) $4\frac{2}{5} - 2\frac{3}{10} = \frac{\text{....}}{5} - \frac{\text{.....}}{10}$

> You can also add or subtract mixed numbers by converting them into improper fractions first.

$= \frac{\text{....}}{10} - \frac{\text{.....}}{10} = \frac{\text{.....}}{10}$

$= \text{..................}$ **(3 marks)**

2 Work out

(a) $1\frac{2}{3} \times 2\frac{3}{10} = \frac{\text{....}}{3} \times \frac{\text{.....}}{10} = \frac{\text{.......}}{\text{.......}} = \text{..................}$ **(3 marks)**

> Replace the ÷ with × and then flip the second fraction over.

(b) $4\frac{2}{3} \div 1\frac{2}{5} = \frac{\text{....}}{3} \div \frac{\text{....}}{5} = \frac{\text{....}}{3} \times \frac{\text{.......}}{\text{.......}} = \frac{\text{.......}}{\text{.......}} = \text{..................}$ **(3 marks)**

3 Work out

(a) $3\frac{1}{2} \times 2\frac{4}{7}$ 　　　　　　(b) $5\frac{1}{3} \div 1\frac{4}{9}$

.... **(3 marks)** 　　　　　　......................... **(3 marks)**

4 A man wins £300 and decides to give it to his three children. He gives $\frac{2}{5}$ of his money to Andrew, $\frac{1}{3}$ to Ben, and the rest to Carla.

> Write 1 as a fraction with the same numerator and denominator. $1 = \frac{15}{15}$

Work out how much money Carla receives.

$\frac{2}{5} + \frac{1}{3} = \frac{\text{....}}{15} + \frac{\text{....}}{15} = \frac{\text{....}}{15}$ 　　$1 - \frac{\text{....}}{15} = \frac{\text{....}}{15}$ 　　$£300 \times \frac{\text{....}}{15} = £ \text{..................}$ **(3 marks)**

5 The diagram shows three identical shapes. $\frac{2}{3}$ of shape A is shaded and $\frac{5}{8}$ of shape C is shaded.

> You will need to use problem-solving skills throughout your exam – **be prepared!**

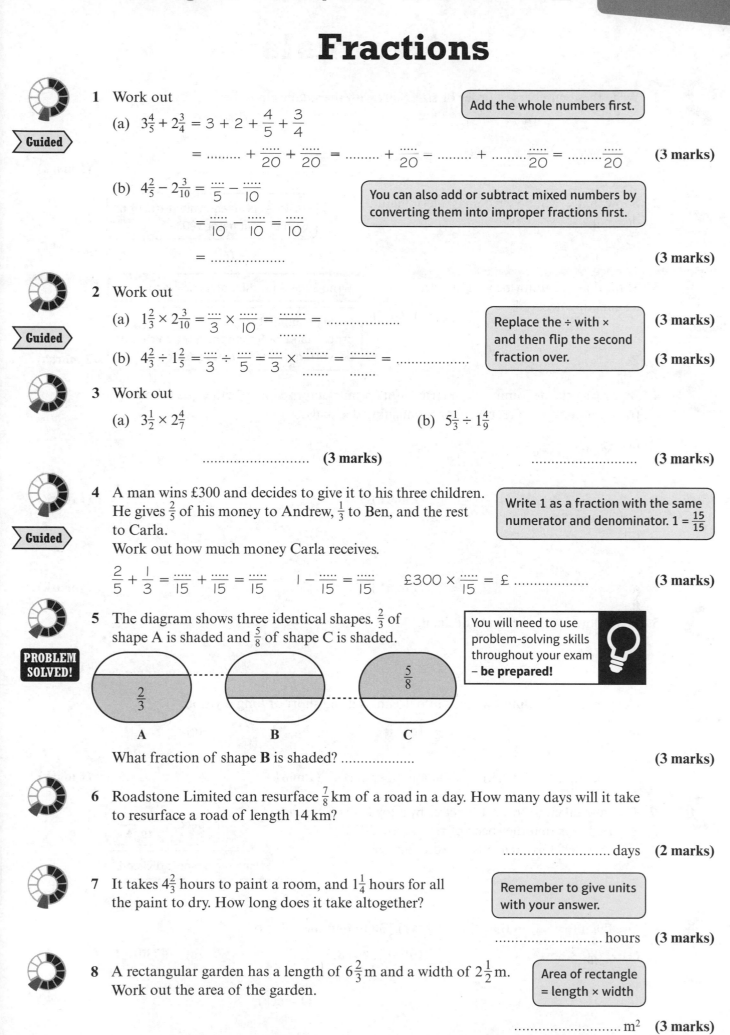

What fraction of shape **B** is shaded? .... **(3 marks)**

6 Roadstone Limited can resurface $\frac{7}{8}$ km of a road in a day. How many days will it take to resurface a road of length 14 km?

.... days **(2 marks)**

7 It takes $4\frac{2}{3}$ hours to paint a room, and $1\frac{1}{4}$ hours for all the paint to dry. How long does it take altogether?

> Remember to give units with your answer.

.... hours **(3 marks)**

8 A rectangular garden has a length of $6\frac{2}{3}$ m and a width of $2\frac{1}{2}$ m. Work out the area of the garden.

> Area of rectangle = length × width

.... m^2 **(3 marks)**

Decimals

1 Write these numbers in order of size. Start with the smallest number.

$\frac{1}{3}$ 0.3 $\frac{18}{50}$ 0.35

.... **(1 mark)**

Guided

2 Show that $\frac{3}{20}$ can be written as a terminating decimal.

$$\frac{3}{20} = \frac{.....}{100} = ...$$

> Write $\frac{3}{20}$ as an equivalent fraction with denominator 100.

(2 marks)

Guided

3 Show that $\frac{7}{30}$ cannot be written as a terminating decimal.

$30 = ... \times .... \times ....$

> Write 30 as a product of its prime factors.

> If the denominator contains a factor other than 2 or 5 then the fraction cannot be written as a terminating decimal.

(2 marks)

4 By writing the denominator in terms of its prime factors, state whether the following fractions convert to recurring or terminating decimals.

(a) $\frac{11}{40}$ (b) $\frac{15}{32}$

(1 mark)   **(1 mark)**

(c) $\frac{22}{39}$ (d) $\frac{9}{42}$

(1 mark)   **(1 mark)**

5 Write down $\frac{2}{11}$ as recurring decimal.

.... **(1 mark)**

6 Convert the following fractions into decimals using short or long division.

(a) $\frac{11}{40}$ (b) $\frac{6}{25}$ (c) $\frac{11}{30}$

.... **(1 mark)**   .... **(1 mark)**   .... **(1 mark)**

7 The time taken to travel 12 metres by a toy car is 5 seconds. Sandeep says that the speed of the car is 2.375 m/s. Is he correct? Give reasons for your answer.

> speed = $\dfrac{\text{distance}}{\text{time}}$

> Use long or short division.

(3 marks)

8 Use the information that $138 \times 85 = 11\,730$ to find the value of

(a) 1380×85 (b) 0.138×8.5 (c) $11\,730 \div 1.38$

.... **(1 mark)**   .... **(1 mark)**   .... **(1 mark)**

Estimation

1 Work out an estimate for the value of

(a) $188 \times 69 \approx 200 \times 70 =$ ....   **(1 mark)**

$\boxed{\text{Round both values to 1 significant figure.}}$

(b) $28.9 \div 4.85 \approx$ .... $\div$ .... $=$ ....   **(1 mark)**

(c) $(51.2)^3 \approx ($.......$)^3 =$ ....   **(1 mark)**

Guided

2 Work out an estimate for the value of $\dfrac{4826}{4.1 \times 9.72}$

$\boxed{\begin{array}{l}\text{1. Round all values to 1 significant figure.}\\ \text{2. Multiply the numbers in the denominator.}\\ \text{3. Cancel if possible, then divide.}\end{array}}$

$\approx \dfrac{5000}{4 \times \text{.......}} = \dfrac{\text{.......}}{\text{.......}} =$ ....

(2 marks)

Guided

3 Work out an estimate for the value of

$\dfrac{8.92 \times 408}{0.506}$

$\boxed{\text{Do not round 0.506 to 1 as this is incorrect.}}$

$\boxed{\begin{array}{l}\text{If you need to divide by a decimal you can}\\ \text{multiply top and bottom by 10 or 100 to}\\ \text{simplify the calculation.}\end{array}}$

.... **(2 marks)**

4 Work out an estimate for the value of $\dfrac{716 \times 5.13}{0.191}$

$\approx \dfrac{700 \times 5}{0.2} = \dfrac{3500}{0.2} = \dfrac{\text{.......}}{2} =$ ....

(2 marks)

Guided

5 Work out an estimate for the value of $\dfrac{29 \times 4.90}{0.204}$

.... **(2 marks)**

6 The radius of a sphere is 6.2 cm.

(a) Work out an estimate for the surface area of the sphere.

$\boxed{\begin{array}{l}\text{You will need to use}\\ \text{problem-solving skills}\\ \text{throughout your exam}\\ \textbf{– be prepared!}\end{array}}$ $\boxed{\begin{array}{l}\text{Surface area of}\\ \text{a sphere} = 4\pi r^2\end{array}}$

PROBLEM SOLVED!

.... cm² **(2 marks)**

(b) Without further calculation, explain whether your method gives you an overestimate or an underestimate for the surface area of the sphere.

.... **(1 mark)**

7 Karl has a field in the shape of a trapezium.

(a) Work out an estimate for the area of the field.

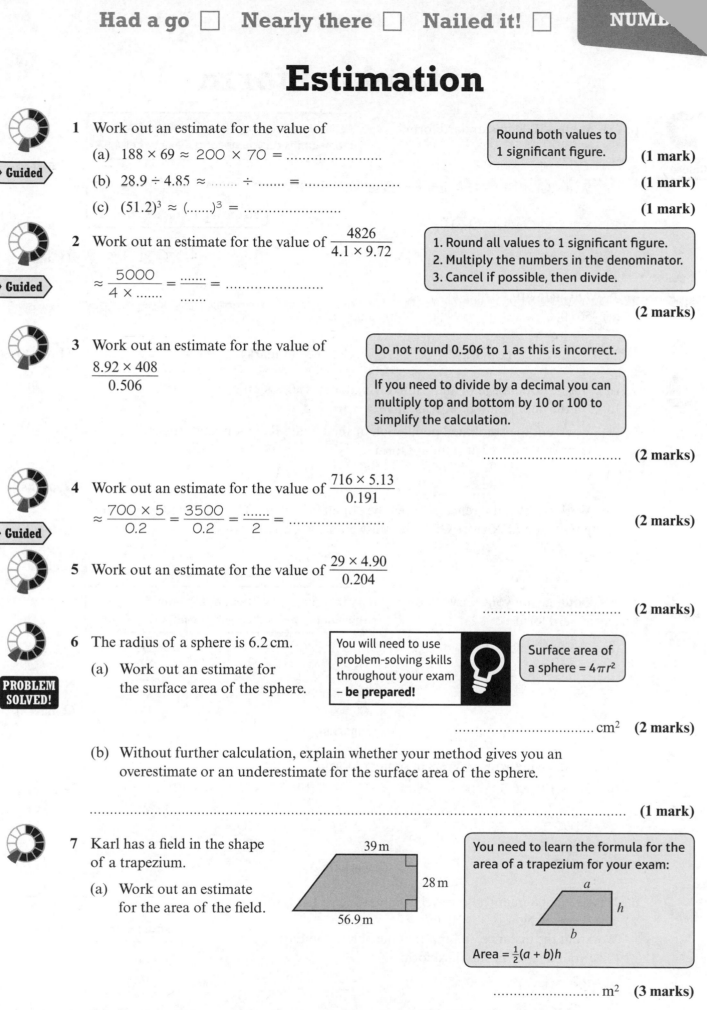

$\boxed{\begin{array}{l}\text{You need to learn the formula for the}\\ \text{area of a trapezium for your exam:}\\ \text{Area} = \frac{1}{2}(a + b)h\end{array}}$

.... m² **(3 marks)**

(b) Is your answer an overestimate or underestimate? Explain your answer.

.... **(1 mark)**

Standard form

1 (a) Write 45 000 in standard form.

> Count decimal places from the right. How many jumps do you need to make to get 4.5?

$45\,000 = 4.5 \times 10^{....}$ **(1 mark)**

(b) Write 3.4×10^{-5} as an ordinary number.

> The power of 10 is negative so the number is less than 1.

$3.4 \times 10^{-5} = 0.0000034$ .... **(1 mark)**

(c) Write 28×10^6 in standard form. $2 \cdot 8 \times 10^7$ .... **(1 mark)**

2 Write in standard form

(a) 567 000

5.67×10^5 .... **(1 mark)**

(b) 0.000 056 7

5.67×10^{-5} .... **(1 mark)**

(c) 567×10^8

5.67×10^{10} .... **(1 mark)**

3 In 2014 the population of the United Kingdom was 6.5×10^7.
In 2014 the population of Russia was 1.4×10^8.

(a) Work out the combined population of the United Kingdom and Russia.
Give your answer in standard form.

$6.5 + 1.4 = 7.9$
$8 + 7 = 15$

7.9×10^{15} .... **(2 marks)**

(b) Work out the difference between the population of the United Kingdom and the population of Russia. Give your answer in standard form.

$(B. \; 6.5 - 1.4 = 5.1$
$7 - 8 = -1$

5.1×10^1 .... **(2 marks)**

4 Work out, giving your answers in standard form

> Try this question without a calculator. Multiply the number parts then add the powers.

(a) $(3 \times 10^6) \times (6 \times 10^{-3})$

$(3 \times 6) \times (10^6 \times 10^{....}) = 1.8 \times 10^4$

$= .... \times 10^{....}$

(2 marks)

(b) $(8 \times 10^6) \div (4 \times 10^{-14})$

$(8 \div ....) \times (10^6 \div 10^{-14}) = 2 \times 10^{20}$

(2 marks)

5 Work out, giving your answers in standard form

(a) $5.1 \times 10^3 + 6.5 \times 10^4$

   5100
+ 65000
   70100

(2 marks)

(b) $7.6 \times 10^5 - 8 \times 10^3$

  760000
−   8000
  68 000

(2 marks)

6 It takes light 8 minutes to travel from the Sun to the Earth.
The speed of light is 3×10^8 m/s.
Work out the distance, in km, from the Sun to the Earth.
Give your answer in standard form.

> Distance (in km) = speed (in km/s) × time (in seconds)

$S = \dfrac{D}{T}$

$3 \times 10^8 = \dfrac{D}{8 \times 60}$

$D = 1.08 \times 10^{11}$ .... **(3 marks)**

Recurring decimals

1 Show that $0.\dot{7}$ can be written as the fraction $\frac{7}{9}$

Let $x = 0.7777777...$

$10x = 7.7777777...$ | Multiply by 10. |

$-$ $x = 0.7777777...$

$.........x = ...$

$x = ...$ **(3 marks)**

2 Prove that the recurring decimal $0.\dot{4}\dot{2}$ has the value $\frac{14}{33}$

Let $x = 0.42424242...$

$100x = 42.42424242...$ | Multiply by 100. |

$-$ $x = 0.42424242...$

$.........x = ...$

$x = ...$ | Cancel down. | **(3 marks)**

3 (a) Show that the recurring decimal $0.8\dot{1}$ can be written as $\frac{9}{11}$

(b) Hence, or otherwise, write the recurring decimal $0.48\dot{1}$ as a fraction.

.... **(3 marks)**   .... **(1 mark)**

4 Work out the recurring decimal $0.6\dot{1}\dot{7}$ as a fraction in its simplest form.

> You will need to use problem-solving skills throughout your exam – **be prepared!**

.... **(3 marks)**

5 Express the recurring decimal $5.23\dot{7}\dot{1}$ as a fraction.

.... **(3 marks)**

6 Prove that the recurring decimal $6.43\dot{2}$ can be written as the fraction $\frac{5789}{900}$

(3 marks)

7 x is an integer such that $1 \leqslant x \leqslant 9$

Prove that $0.\dot{0}\dot{x} = \frac{x}{99}$

(3 marks)

Upper and lower bounds

Guided

1   The mass of a bag of cement is 20 kg, correct to the nearest kg. | What is half of 1 kg? |

    (a) Write down the smallest possible mass of the bag of cement.

    (b) Write down the largest possible mass of the bag of cement.

    $20 - 0.5 =$ .... kg   **(1 mark)** $20 +$ .... $=$ .... kg   **(1 mark)**

Guided

2   The length of a piece of string is 52.3 cm, correct to 1 decimal place.

    (a) Write down the greatest possible length of the piece of string.

    (b) Write down the least possible length of the piece of string.

    $52.3 + 0.05 =$ .... cm   **(1 mark)** $52.3 -$ .... $=$ .... cm   **(1 mark)**

Guided

3   The kinetic energy, in joules (J), of a moving object is calculated using the formula

        kinetic energy $= \frac{1}{2}mv^2$

| Use the rules of multiplication when finding upper and lower bounds. |

    The mass of the object is 2.6 kg, to the nearest tenth of a kilogram.

    The velocity of the object is 32.7 m/s, correct to 3 significant figures.

    Find the lower bound and the upper bound of the kinetic energy, in joules, of the object.

    Give your answer correct to 3 significant figures.

    Lower bound of mass = ...   Upper bound of mass = ...

    Lower bound of velocity = ...   Upper bound of velocity = ...

    $\frac{1}{2}mv^2 =$ .... $=$ .... J $\frac{1}{2}mv^2 =$ .... $=$ .... J   **(3 marks)**

4   An experiment is carried out to measure the density of rolled lead, in g/cm³.

    The mass of the rolled lead is 572 grams, correct to the nearest gram.

    The volume of the rolled lead is 50.2 cm³, correct to 3 significant figures.

    Use the formula density $= \dfrac{\text{mass}}{\text{volume}}$ to find the range of possible values for the density of rolled lead, in g/cm³.

    Give your answers correct to 4 significant figures.

    .... **(3 marks)**

5   A ball is thrown vertically upwards with a speed v metres per second.

    The height, H metres, to which it rises is given by the formula

        $H = \dfrac{v^2}{2g}$

    where g m/s² is the acceleration due to gravity.

    $v = 35.3$ m/s correct to 3 significant figures and $g = 9.8$ m/s² correct to 2 significant figures.

    Calculate the upper bound of H. Give your answer correct to 3 significant figures.

    .... **(2 marks)**

Accuracy and error

1 The length of a rod is measured as 33.4 cm. Complete the inequality for the length.

> **Guided**

     33.3 .... cm ≤ length of rod < .... cm   **(2 marks)**

2 A bottle of lemonade is labelled with a volume of 1.75 litres.
Complete the inequality for the volume.

     .... litres ≤ volume of lemonade < .... litres   **(2 marks)**

3 The area of a circle is given as 132 cm², correct to 3 significant figures.
Find the radius of the circle to an appropriate degree of accuracy.
You must explain why your answer is to an appropriate degree of accuracy.

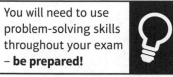

Find the upper and lower bound of the radius.

> **Guided**

> **PROBLEM SOLVED!**

$UB = \sqrt{\dfrac{\text{........}}{\pi}} = $ .... $LB = $ ....

Radius = ...

You will need to use problem-solving skills throughout your exam – **be prepared!**

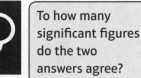

To how many significant figures do the two answers agree?

.... **(4 marks)**

4 A crane uses a cable with a breaking strain of 2800 kg measured to 2 significant figures.
It is used to lift pallets with a mass of 50 kg measured to 2 significant figures.
What is the greatest number of pallets that can safely be lifted at one time without the cable breaking?

.... **(4 marks)**

5 Ravina measured her handspan with a ruler and found it was 15 cm to the nearest centimetre. Anjali measured her handspan more accurately and found it was 148 mm to the nearest millimetre.
Use the upper and lower bounds of the measurements to show whether it is possible that Anjali's handspan is larger than Ravina's.

(4 marks)

6 An object is projected vertically upwards.
The energy, E, of the object is 18.3 J, correct to 3 significant figures.
The mass, m, of the object is 3.8 kg, correct to 1 decimal place.

The velocity, v, of the object is given by the formula $v = \sqrt{\dfrac{2E}{m}}$

(a) Find the upper and lower bounds of the velocity, in m/s.
Give your answers correct to 4 significant figures.

(4 marks)

(b) Use your answers to part (a) to write down the value of v to a suitable degree of accuracy.
You must explain your answer.

.... **(1 mark)**

Surds 1

1 Express the following surds in the form $a\sqrt{b}$, where a and b are integers.

(a) $\sqrt{12} = \sqrt{........} \times \sqrt{3} = ...\sqrt{3}$ **(1 mark)**

> Think of two numbers which, when multiplied, give 12, one of which is a square number.

(b) $\sqrt{20} = \sqrt{4} \times \sqrt{........} = ...$ **(1 mark)**

(c) $\sqrt{48} = \sqrt{........} \times \sqrt{........} = ...$ **(1 mark)**

(d) $\sqrt{3} \times \sqrt{27}$ .... **(2 marks)**

(e) $\sqrt{98} - \sqrt{18}$ .... **(2 marks)**

(f) $5\sqrt{28} - \sqrt{63}$ .... **(2 marks)**

2 Rationalise the denominator.

(a) $\dfrac{3}{\sqrt{5}}$

$\dfrac{3}{\sqrt{5}} \times \dfrac{\sqrt{5}}{\sqrt{5}} =$ .... **(1 mark)**

(b) $\dfrac{2}{\sqrt{6}}$

$\dfrac{2}{\sqrt{6}} \times \dfrac{\sqrt{....}}{\sqrt{....}} =$ .... = ... **(1 mark)**

(c) $\dfrac{2}{\sqrt{8}}$

.... **(1 mark)**

(d) $\dfrac{21}{\sqrt{7}}$

.... **(1 mark)**

(e) $\dfrac{1 + \sqrt{3}}{\sqrt{12}}$

.... **(2 marks)**

(f) $\dfrac{\sqrt{18} + 10}{\sqrt{2}}$

.... **(2 marks)**

3 Solve the following equations where x is an integer.

(a) $\sqrt{45} \times x = \sqrt{180}$

$x =$ .... **(3 marks)**

(b) $\dfrac{\sqrt{x} \times \sqrt{24}}{2\sqrt{3}} = \sqrt{20}$

$x =$ .... **(4 marks)**

4 Find the length of the side labelled x in the form $a\sqrt{b}$, where a and b are integers.

$x =$ .... **(3 marks)**

5 Show that $\dfrac{1}{1 + \dfrac{1}{\sqrt{3}}}$ can be written as $\dfrac{3 - \sqrt{3}}{2}$

(3 marks)

Counting strategies

1   Emily has four tiles.

W	X	Y	Z

Emily chooses two of these tiles.
Write down all the possible combinations she can get.

(2 marks)

2   Asha, Bev, Chloe and Dan are playing in a competition.
Each player must play each other once.
How many games will be played in total?

> **Guided**

Label Asha, Bev, Chloe and Dan as A, B, C and D respectively.

Remember (A, B) is the same as (B, A).

(A, ...) (A, ...) (A, ...) (B, ...) .... **(2 marks)**

3   Craig has a black shirt (BS), a white shirt (WS) and a pink shirt (PS).
He also has a yellow tie (YT), a red tie (RT) and an orange tie (OT).
Craig picks a shirt and a tie combination at random.
How many combinations can Craig choose?

.... **(2 marks)**

4   Jack has a code for his money box. The code consists of two digits followed by three
letters. The digits and the letters can be repeated. The digits are the numbers 0 to 9.
Jack says that there are more than one million different possible codes.
Is he correct? You must show your working.

> **Guided**

How many digits are there?

How many letters are there?

digit   digit   letter   letter   letter

10 × .... × 26 × .... × .... = ...

Jack is .... **(2 marks)**

5   The diagrams show keypads for two different types of alarm.
Each keypad has a four-key code.

> **PROBLEM SOLVED!**

You will need to use problem-solving skills throughout your exam – **be prepared!**

Premier alarm keypad

Classic alarm keypad

1 2 3 4 5
6 7 8 9 0

1 2 3 4 5
6 7 8 9 0
A B C

(a) How many different codes are possible for the

   (i) Classic alarm keypad   (ii) Premier alarm keypad?

.... **(2 marks)**

(b) The Premier alarm keypad is then programmed so that the four-key code must
start with two letters, followed by two digits.
Show that there are fewer than 1000 codes possible.

(2 marks)

13

Problem-solving practice 1

1 A machine makes 48 bolts every hour.
The machine makes bolts for $7\frac{1}{2}$ hours each day, on 5 days of the week.
The bolts are packed into boxes. Each box holds 30 bolts.
How many boxes are needed for all the bolts made each week?

.... boxes   **(4 marks)**

2 Ethan bought some food for a party. He is going to make some hot dogs.
He needs a bread roll, a sausage and a sachet of ketchup for each hot dog.
There are 40 bread rolls in a pack, 24 sausages in a tray and 15 sachets of ketchup in a packet.
Ethan buys exactly the same number of bread rolls, sausages and sachets of ketchup.

(a) What is the smallest number of packs of bread rolls, trays of sausages and packets of ketchup Ethan could have bought?

.... packs of bread rolls

.... trays of sausages

.... packets of ketchup   **(3 marks)**

(b) How many hot dogs can he make?

.... **(1 mark)**

3 (a) A code for a mobile phone is made up of three digits.
The digits use the numbers 0 to 9. The digits can be repeated.
How many different possible codes are there?

.... **(1 mark)**

(b) If the digits cannot be repeated, how many different possible codes are there?

.... **(2 marks)**

A different code has x digits. The digits use the numbers 1 to 8. The digits can be repeated.
There are about 260 000 different possible codes.

(c) Work out the value of x. You must show all your working.

$x =$ .... **(2 marks)**

Problem-solving practice 2

4 One sodium atom has a mass of 3.82×10^{-23} grams.

(a) Work out an estimate for the number of sodium atoms in 1 kg of sodium.

.... **(3 marks)**

(b) Is your answer to part (a) an underestimate or an overestimate?
Give a reason for your answer.

....

.... **(1 mark)**

5 $t = \dfrac{d}{2\sqrt{f}}$

$d = 9.82$ correct to 3 significant figures
$f = 2.46$ correct to 3 significant figures
By considering bounds, work out the value of t to a suitable degree of accuracy.
Give a reason for your answer.

(5 marks)

6 A large rectangular piece of card is $(5 + \sqrt{8})$ cm long and $(\sqrt{2} + 2)$ cm wide.
A small rectangle $\sqrt{6}$ cm long and $\sqrt{3}$ cm wide is cut out of the piece of card.

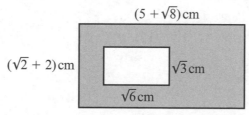

$(5 + \sqrt{8})$ cm

$(\sqrt{2} + 2)$ cm

$\sqrt{3}$ cm

$\sqrt{6}$ cm

Work out the shaded area, in cm². Give your answer in the form $a\sqrt{2} + b$ where
a and b are integers.

.... cm² **(4 marks)**

Algebraic expressions

Guided

Simplify fully

(a) $m \times m \times m$

$m^{\underline{3}}$ **(1 mark)**

(b) $d \times d \times d \times d$

$d^{\underline{4}}$ **(1 mark)**

(c) $e \times e \times e \times e \times e$

$e^{\underline{5}}$ **(1 mark)**

2 Simplify

Guided

(a) $x^4 \times x^7$

$x^{\underline{4} + \underline{7}} = x^{\underline{11}}$ **(2 marks)**

(b) $y^7 \div y^2$

$y^{\underline{7} - \underline{2}} = y^{\underline{5}}$ **(2 marks)**

(c) $t^5 \times t^6 \div t^7$

$4 \quad T^4$ **(2 marks)**

3 Simplify fully

Guided

(a) $(x^3)^2$

$x^{\underline{3} \times \underline{2}} = x^{\underline{6}}$ **(1 mark)**

(b) $(y^5)^3$

y^{15} **(2 marks)**

(c) $(t^3)^7$

T^{21} **(2 marks)**

4 Simplify fully

Guided

(a) $\dfrac{x^3 \times x^4}{x^2}$

$\dfrac{x^{\underline{3} + \underline{4}}}{x^2} = x^{\underline{7} - \underline{2}}$

$= x^{\underline{5}}$ **(1 mark)**

(b) $\dfrac{y^{14}}{y^3 \times y^2}$

y^9 **(2 marks)**

(c) $\left(\dfrac{t^7}{t^4}\right)^2$

y^6 **(2 marks)**

5 Simplify fully

(a) $7xy^3 \times 4x^2y^4$

$28x^3y^7$ **(2 marks)**

(b) $\dfrac{16x^4y^3}{8xy^2}$

$2x^3y^1$ **(2 marks)**

(c) $(3x^2y^5z^3)^4$

$81x^8 \quad 20 \quad 12$ **(2 marks)**

6 Simplify fully

(a) $(25x^6)^{\frac{1}{2}}$

$5x^3$ **(2 marks)**

(b) $(16x^3y^4)^{\frac{3}{2}}$

$64x^{\frac{9}{2}}y^6$ **(2 marks)**

(c) $(81x^5y^3)^{\frac{1}{4}}$

$3x^{\frac{5}{4}}y^{\frac{3}{4}}$ **(2 marks)**

7 Simplify fully

(a) $\left(\dfrac{1}{3x^4}\right)^{-2}$

9^{-8} **(2 marks)**

(b) $\left(\dfrac{25}{64x^4y^{10}}\right)^{-\frac{1}{2}}$

$\dfrac{8}{5}x^{-2}y^{-5}$ **(2 marks)**

(c) $\left(\dfrac{27}{64x^3y^9}\right)^{-\frac{2}{3}}$

$\dfrac{16}{9}x^{-2}y^{-6}$ **(2 marks)**

Expanding brackets

1 Expand and simplify

(a) $(x + 3)(x + 4)$ (b) $(x + 5)(x - 3)$ (c) $(x - 2)(x - 6)$

$x(x + 4) + 3(x + 4)$ $x(x - 3) + 5(x - 3)$

$= x^{....} + ...x + ...x + 12$ $= x^{....} - ....x + ...x - ....$

$= x^{....} + ...x + 12$ $= x^{....} + ...x - ....$

(2 marks)   **(2 marks)**   **(2 marks)**

2 Expand and simplify

(a) $(x + 3)^2$ (b) $(x - 4)^2$ (c) $(2x + 1)^2$

$(x + 3)(x + 3)$

$= x(x + 3) + 3(x + 3)$

$= x^{....} + ...x + ...x + ...$

$= x^{....} + ...x + ...$

(2 marks)   **(2 marks)**   **(2 marks)**

3 Expand and simplify | Multiply out the brackets first and then multiply this expression by x.

(a) $x(x + 3)(x + 5)$ (b) $x(x - 2)(x + 4)$ (c) $x(x - 3)(x - 7)$

$x(x + 3)(x + 5)$

$= x(x^{....} + ...x + ...x + ...)$

$= x(x^{....} + ...x + ...)$

$= x^{....} + ...x^{....} + ...x$

(2 marks)   .... **(2 marks)**   .... **(2 marks)**

4 Expand and simplify

(a) $(x + 3)^3$ (b) $(x - 4)^3$ (c) $(2x + 1)^3$

.... **(2 marks)**   .... **(2 marks)**   .... **(2 marks)**

5 The diagram shows a cuboid with a length of $x + 2$, a width of $x - 3$, and a height of $x + 4$. All measurements are in cm.

(a) Write down an expression, in terms of x, for the total surface area.

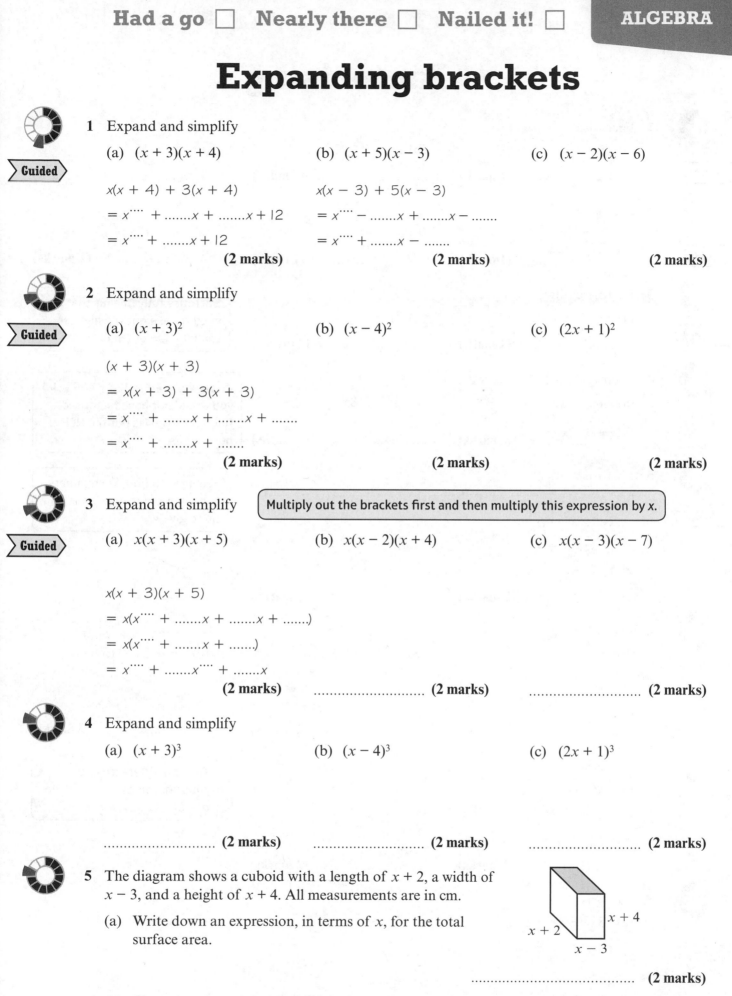

.... **(2 marks)**

(b) Show that the volume of the cuboid can be written in the form $ax^3 + bx^2 + cx + d$ cm^3.

(2 marks)

Factorising

1 Factorise

(a) $3x + 6$

(b) $2p - 6$

(c) $5y - 15$

= 3(....... + ...) **(1 mark)**

.... **(1 mark)**

= 5(....... − ....) **(1 mark)**

2 Factorise

(a) $x^2 + 6x$

(b) $x^2 + 4x$

(c) $x^2 - 12x$

= x(....... + ...) **(1 mark)**

.... **(1 mark)**

= x(....... − ....) **(2 marks)**

3 Factorise fully

(a) $3p^2 + 6p$

(b) $8y^2 - 24y$

= 3p(....... + ...) **(1 mark)**

.... **(2 marks)**

> 'Factorise fully' means that you need to take out the highest common factor (HCF).

4 Factorise fully

(a) $4d^2 + 12d$

(b) $6x^2 - 18x$

.... **(2 marks)**

.... **(2 marks)**

> If you wrote $4d^2 + 12d = 4(d^2 - 3d)$ you would not have factorised fully, because 4 is not the HCF of both terms.

5 Factorise

(a) $x^2 + 4x + 3$

(b) $x^2 + 11x + 10$

..... × ..... = +3

..... × ..... = +10

..... + ... = +4

..... + ... = +11

$x^2 + 4x + 3 = (x + ...)(x + ...)$

$x^2 + 11x + 10 = (x ....)(x ....)$

(2 marks)

(2 marks)

> You need to find two numbers that multiply to give 3 and add up to give 4.

6 Factorise

(a) $x^2 + 6x - 7$

(b) $x^2 + 4x - 5$

(c) $x^2 - 2x - 15$

.... **(2 marks)**

.... **(2 marks)**

.... **(2 marks)**

7 Factorise

(a) $x^2 - 9$

(b) $x^2 - 144$

$a = x, b = 3$

$x^2 - 9 = (x + ...)(x - .....)$

(2 marks)

.... **(2 marks)**

> This is a difference of two squares. You can use the rule $a^2 - b^2 = (a + b)(a - b)$.

8 Factorise

(a) $3x^2 - 7x + 2$

(b) $2x^2 - x - 3$

(c) $3x^2 - 16x - 12$

.... **(2 marks)**

.... **(2 marks)**

.... **(2 marks)**

Linear equations 1

1 Solve

(a) $3x + 1 = 13$ (b) $5x - 3 = 27$ (c) $26 = 7q - 9$

$3x + 1 = 13$ $(- 1)$

$3x = 13 - 1$

$3x = \text{.......}$ $(÷ 3)$

    $x = \text{.......}$ **(1 mark)** $x = \text{.......}$ **(1 mark)** $q = \text{.......}$ **(1 mark)**

(d) $12x + 18 = 66$ (e) $\dfrac{t}{6} - 7 = 3$ (f) $\dfrac{d}{3} + 2 = -4$

    $x = \text{.......}$ **(1 mark)** $t = \text{.......}$ **(1 mark)** $d = \text{.......}$ **(1 mark)**

2 Solve

> Multiply out the brackets.

(a) $3(3x + 5) = 42$ (b) $5(2x + 3) = 35$ (c) $5(x - 3) = -25$

$9x + \text{.......} = 42$

$9x = 42 - \text{.......}$

$9x = \text{.......}$ $(÷ 9)$

    $x = \text{.......}$ **(2 marks)** $x = \text{.......}$ **(2 marks)** $x = \text{.......}$ **(2 marks)**

(d) $4(5x + 7) = 16$ (e) $3(4x + 13) = 51$ (f) $3(10 - 4x) = 45$

    $x = \text{.......}$ **(2 marks)** $x = \text{.......}$ **(2 marks)** $x = \text{.......}$ **(2 marks)**

3 Solve

> Collect all the x terms on one side.

(a) $2x + 3 = x + 7$ (b) $7y + 15 = 4y - 6$ (c) $4t - 6 = 2t + 18$

    $x = \text{.......}$ **(2 marks)** $y = \text{.......}$ **(2 marks)** $t = \text{.......}$ **(2 marks)**

(d) $2(x + 3) = x + 10$ (e) $5(x - 4) = 3(x + 2)$ (f) $3(2y - 4) = 2(6 - 3y)$

    $x = \text{.......}$ **(2 marks)** $x = \text{.......}$ **(2 marks)** $y = \text{.......}$ **(2 marks)**

4 Carl buys 8 bags of marbles. Each bag contains m marbles. He plays his friend and wins another 7 marbles. When Carl gets home, he counts his marbles and finds that he has 103 marbles altogether. Calculate the value of m. You must show all of your working.

> You can write an equation involving m. Carl starts with $8m$ marbles. He then adds 7 to get 103.

    $m = \text{..........................}$ **(3 marks)**

Linear equations 2

1 Solve

(a) $\dfrac{3x + 10}{2} = 12$ (b) $\dfrac{2x + 7}{5} = 3$ (c) $4 = \dfrac{5x - 3}{3}$

$$\dfrac{3x + 10}{2} = 12 \qquad (\times\ 2)$$

$$\dfrac{2(3x + 10)}{2} = 12 \times 2$$

$$3x + 10 = \text{.....} \qquad (-\ 10)$$

$$3x = \text{.....} - \text{.....}$$

$$3x = \text{.....} \qquad (\div\ 3)$$

$$x = \text{.......}$$ **(3 marks)** $x = \text{.......}$ **(3 marks)** $x = \text{.......}$ **(3 marks)**

2 Solve

(a) $\dfrac{6 - 2x}{4} = 2 - x$ (b) $\dfrac{5x}{3} - 6 = x + 2$

> Place brackets around $2 - x$.

$$\dfrac{6 - 2x}{4} = 2 - x \qquad (\times\ 4)$$

$$\dfrac{4(6 - 2x)}{4} = 4(2 - x)$$

> Multiply out the brackets.

$$6 - 2x = 4(2 - x)$$

$$6 - 2x = \text{.....} - \text{.....}x$$

> Collect all the x terms on one side.

$$\text{.....}x - 2x = \text{.....} - 6$$

$$\text{.....}x = \text{..............}$$ **(3 marks)** $x = \text{..............}$ **(3 marks)**

3 Solve

(a) $\dfrac{x - 5}{5} - \dfrac{4 - x}{3} = 5$ (b) $\dfrac{3x - 1}{2} - \dfrac{2(4 + 3x)}{13} = 2$

$$\dfrac{3(x - 5)}{5 \times 3} - \dfrac{5(4 - x)}{3 \times 5} = 5$$

$$\dfrac{3x - \text{.....} - \text{.....} + \text{.....}x}{15} = 5 \qquad (\times\ 15)$$

$$\text{.....}x - \text{.....} = 5 \times 15$$

$$\text{.....}x - \text{.....} = 75$$

$$\text{.....}x = \text{.....}$$

$$\text{.....}x = \text{..............}$$ **(3 marks)** $x = \text{..............}$ **(3 marks)**

4 ABC is a triangle. Angle A is $(4x - 25)°$ and angle B is $(x + 3)°$.
The size of angle A is three times the size of angle B.
Work out the value of x.

> Remember to set up an equation.

$$x = \text{..............}$$ **(3 marks)**

Formulae

1 Using the formula $y = 3x - 7$, find the value of y when

(a) $x = 5$ | Substitute $x = 5$ into the formula. | (b) $x = -4$

$y = 3(5) - 7$

$\quad = \ldots - 7 = \ldots\ldots\ldots\ldots\ldots$ **(2 marks)**

$y = 3(\ldots) - 7$

$\quad = \ldots - 7 = \ldots\ldots\ldots\ldots\ldots$ **(2 marks)**

2 The value of y can be found by using the formula $y = 2x^2 + 5$. Work out the value of y when

(a) $x = 3$ (b) $x = \frac{5}{2}$

$y = 2(\ldots)^2 + 5$

$\quad = \ldots + 5 = \ldots\ldots\ldots\ldots\ldots$ **(2 marks)** $y = \ldots\ldots\ldots\ldots\ldots$ **(3 marks)**

3 Give answers to parts (a) to (c) correct to 3 significant figures.

(a) Using the formula $V = \frac{1}{3}\pi r^2 h$, work out the value of V when $r = 24$ and $h = 5.8$.

$V = \frac{1}{3}\pi(\ldots)^2 \times \ldots = \ldots\ldots\ldots\ldots\ldots$ **(2 marks)**

(b) Using the formula $s = ut + \frac{1}{2}at^2$, work out the value of s when $u = -4$, $t = 6$ and $a = -9.8$.

$s = \ldots\ldots\ldots\ldots\ldots$ **(3 marks)**

(c) Using the formula $T = 2\pi\sqrt{\frac{L}{g}}$, work out the value of T when $L = 19.4$ and $g = 9.8$.

$T = \ldots\ldots\ldots\ldots\ldots$ **(2 marks)**

(d) Using the formula $T = \dfrac{2Mmg}{M + m}$, work out the value of T when $M = 3.6$, $m = 1.4$ and $g = 9.8$.
   Give your answer correct to 2 significant figures.

$T = \ldots\ldots\ldots\ldots\ldots$ **(2 marks)**

4 The number, N, of plants per hectare is given by the formula $N = \dfrac{12\,000}{dx}$, where d is the distance between the rows of plants in metres and x is the spacing between the plants in metres. Given that $d = 0.85$ and $x = 0.54$, work out the value of N.
Give your answer correct to 2 significant figures.

$N = \ldots\ldots\ldots\ldots\ldots$ **(3 marks)**

5 The surface area of a cylinder is given by the formula

$A = 2\pi r^2 + 2\pi rh$

The volume of a cylinder is given by the formula $V = \pi r^2 h$
Find a formula for A in terms of r and V.

| You will need to use problem-solving skills throughout your exam – **be prepared!** |

$A = \ldots\ldots\ldots\ldots\ldots$ **(3 marks)**

PROBLEM SOLVED!

Arithmetic sequences

1 Here are some sequences. Find an expression for the *n*th term of each linear sequence.

(a) 5   9   13   17

+4 +4 +4

> Work backwards to find the zero term of the sequence. You need to subtract the difference from the first term. Then *n*th term = difference × *n* + zero term.

*n*th term = 4*n* .... **(2 marks)**

(b) 2   5   8   11

.... **(2 marks)**

(c) 2   9   16   23

.... **(2 marks)**

2 Here are the first five terms of a linear sequence.

4   7   10   13   16

Find an expression, in terms of *n*, for the *n*th term of the linear sequence.

.... **(2 marks)**

3 Here are some patterns made from sticks.

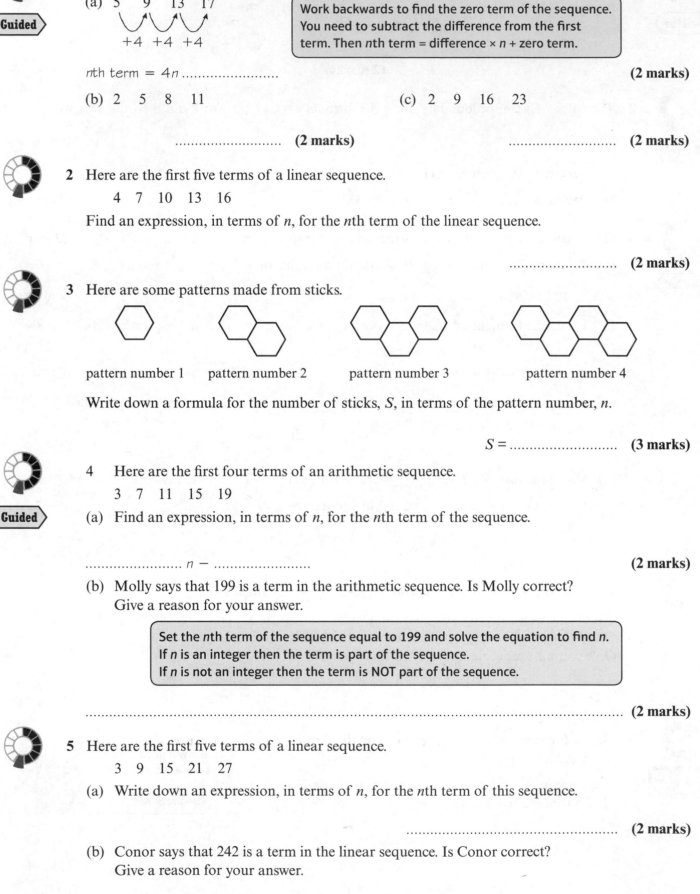

pattern number 1   pattern number 2   pattern number 3   pattern number 4

Write down a formula for the number of sticks, *S*, in terms of the pattern number, *n*.

S = ... **(3 marks)**

4 Here are the first four terms of an arithmetic sequence.

3   7   11   15   19

(a) Find an expression, in terms of *n*, for the *n*th term of the sequence.

.... *n* − .... **(2 marks)**

(b) Molly says that 199 is a term in the arithmetic sequence. Is Molly correct?
Give a reason for your answer.

> Set the *n*th term of the sequence equal to 199 and solve the equation to find *n*.
> If *n* is an integer then the term is part of the sequence.
> If *n* is not an integer then the term is NOT part of the sequence.

.... **(2 marks)**

5 Here are the first five terms of a linear sequence.

3   9   15   21   27

(a) Write down an expression, in terms of *n*, for the *n*th term of this sequence.

.... **(2 marks)**

(b) Conor says that 242 is a term in the linear sequence. Is Conor correct?
Give a reason for your answer.

.... **(2 marks)**

Solving sequence problems

Guided

1. The nth term of an arithmetic sequence is given by $an + b$, where a and b are integers. The 4th term is 11 and the 7th term is 20. Work out the values of a and b.

> Set up two equations and solve them simultaneously.

$11 = a \times \dots\dots\dots + b \qquad$ and $\qquad 20 = a \times \dots\dots\dots + b$

.... **(2 marks)**

2. The rule for finding the next term in a sequence is | Multiply by k then add 3 |

The second term of the sequence is 21 and the third term is 45.

(a) Work out the first term of the sequence.

.... **(3 marks)**

(b) Write down an expression for u_{n+1} in terms of u_n.

.... **(1 mark)**

3. The rule for finding the next term in a sequence is | Add 6 then divide by p |
The second term is 15 and the third term is 7.
Work out the first term of the sequence.

.... **(4 marks)**

4. Here are the first seven terms of a Fibonacci sequence.

$$1 \quad 1 \quad 2 \quad 3 \quad 5 \quad 8 \quad 13$$

To find the next term in the sequence, add the two previous terms.

(a) Work out the tenth term of the Fibonacci sequence.

.... **(1 mark)**

The first three terms of a different Fibonacci sequence are
$$x \quad y \quad x + y$$

(b) Work out the fifth term of the sequence.

.... **(2 marks)**

The third term of the sequence is 11 and the fifth term of the sequence is 28.

(c) Work out the value of x and the value of y.

$x = \dots\dots\dots\dots\dots$ and $y = \dots\dots\dots\dots\dots$ **(3 marks)**

5. $3 \quad 3\sqrt{3} \quad 9 \quad 9\sqrt{3} \quad 27$
Write down the next two terms of the sequence.

.... **(2 marks)**

Quadratic sequences

A quadratic sequence is given by $u_n = n^2 + 2n - 1$

> Substitute $n = 1, 2, 3, 4, 5$ and 6 into the formula.

Write down the first six terms of the sequence.

$u_1 = 1^2 + 2(1) - 1 = \underline{2}$ $u_2 = (\underline{2})^2 + 2(\underline{2}) - 1 = \underline{7}$

$u_3 = (\underline{3})^2 + 2(\underline{3}) - 1 = \underline{14}$ $u_4 = (\underline{4})^2 + 2(\underline{4}) - 1 = \underline{23}$

$u_5 = (\underline{5})^2 + 2(\underline{5}) - 1 = \underline{34}$ $u_6 = (\underline{6})^2 + 2(\underline{6}) - 1 = \underline{47}$ **(2 marks)**

2 Write down the formula for the nth term for each of these quadratic sequences.

(a) 3   6   11   18   27   38

+3 +5

+2

$a = 1$ $c = 2$

$b = 0$

$\underline{n^2 + 2}$ **(3 marks)**

(b) 3   13   27   45   67   93

+10 +14

+4

$c = -9$

$a = 2$ $b = 4$ cubic $= 0$

$3a + b = 10$

$6 + b = 10$ $2n^2 + 4n - 9$ **(3 marks)**

(c) 4   10   20   34   52   74

+6 +10

+4

$a = 2$

$b = 0$

$c = 2$

$\underline{2n^2 + 2}$ **(3 marks)**

(d) 2   9   22   41   66   97

+7 +13

+6

$a = 3$

$b = -2$

$c = 1$

$\underline{3n^2 - 2n + 1}$ **(3 marks)**

(e) 2   12   26   44   66   92

+10 +14

+4

$a = 2$

$b = 4$ $c = -4$

$\underline{2n^2 + 4n - 4}$ **(3 marks)**

(f) 1   5   15   31   53   81

+4 +16

+6 $c = 3$

$a = 3$

$b = -5$

$\underline{3n^2 - 5n + 3}$ **(3 marks)**

3 Here are some patterns made from square slabs.

(a) Write down an expression, in terms of n, for the nth term of this sequence.

$9 + b = 6$ $a = 3$

$b = -3$

+6   +12

+6 $3n^2 - 3n + 2$ **(3 marks)**

(b) Jane says that 75 is a term in the quadratic sequence. Is Jane correct? Give a reason for your answer.

$3(5)^2 - 3(5) + 2 = 62 < 75$ **(1 mark)**

Straight-line graphs 1

Guided

1 On the grid draw the graph of $x + y = 5$ for the values of x from -2 to 5.

> First draw a table of values. The question tells you to use 'values of x from -2 to 5'. Next work out the values of y.

x	-2	-1	0	1	2	3	4	5
y		6						

(2 marks)

Guided

2 Find the equation of the straight line below.

> Draw a triangle on the graph and use it to find the gradient.

$$\text{Gradient} = \frac{\text{distance up}}{\text{distance across}} = \dots\dots\dots\dots$$

> Use $y = mx + c$ to find the equation of the line, where m is the gradient and c is the y-intercept.

 (2 marks)

3 Find the equation of the straight line.

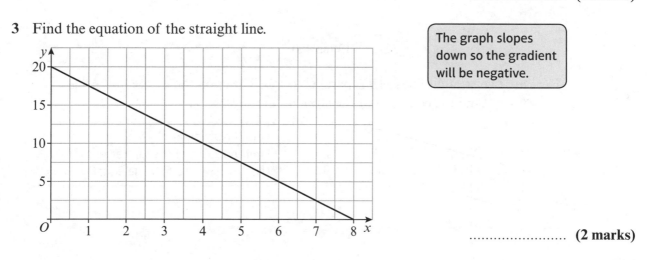

> The graph slopes down so the gradient will be negative.

$\dots\dots\dots\dots$ **(2 marks)**

Straight-line graphs 2

Guided

1 Find the equation of the straight line with

(a) gradient 3, passing through the point (2, 5)

> Substitute the value of the gradient into $y = mx + c$. Then substitute the x- and y-values given into your equation. Solve the equation to find c. Remember to write your completed equation at the end.

$y = 3x + c$

$5 = ... \times ...... + c$

$c = ...$

$y = ...x ....$ **(2 marks)**

(b) gradient −2, passing through the point (3, 6)

(c) gradient 4, passing through the point (−2, 7)

.... **(2 marks)**   .... **(2 marks)**

Guided

2 Find the equation of the straight line which passes through the points

(a) (3, 2) and (5, 6)

> Work out the gradient and then substitute x- and y-values for one of the points into the equation $y = mx + c$ to work out c.

$m = \dfrac{...... - ......}{...... - ......} = ...$

$2 = ... \times 3 + c$

$c = ...$

$y = ...$ **(3 marks)**

(b) (2, 5) and (4, 9)

(c) (−1, −2) and (−4, −8)

.... **(3 marks)**   .... **(3 marks)**

3 Here are two straight lines. *AB* is parallel to *PQ*. The equation of the line *AB* is $y = 4x + 1$. Find the equation of the straight line *PQ*.

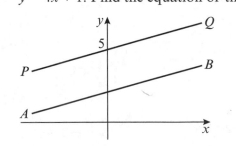

> Find the gradient of *AB*.
> Find the value of the y-intercept for *PQ*.
> Use $y = mx + c$ to find the equation of *PQ*.

.... **(2 marks)**

4 A straight line passes through the points with coordinates (−1, −1), (3, 15) and (k, 23).
Work out the value of *k*. You must show all your working.

> First work out the gradient of the line using the given points.

$k = ...$ **(4 marks)**

Parallel and perpendicular

1 Here are the equations of five straight lines.

P: $y = 2x + 7$ **Q**: $y = -2x + 7$ **R**: $y = x + 7$ **S**: $y = -\frac{1}{2}x + 8$ **T**: $y = \frac{1}{2}x + 4$

(a) Write down the letter of the line that is parallel to $y = x + 4$

> When two lines are parallel their gradients are the same.

.... **(1 mark)**

(b) Write down the letter of the line that is perpendicular to $y = 2x - 3$

> When the gradient of a line is m, the gradient of a perpendicular line is $-\frac{1}{m}$.

.... **(1 mark)**

2 (a) A straight line **L** is parallel to $y = 3x - 4$ and passes through the point $(4, 5)$. Find the equation of line **L**.

Guided

$m = $ ....

$...... = ...(......) + c$

> Compare the straight line with $y = mx + c$ to find the value of m.

Rearranging for c

$c = $ ....

Hence, $y = $ ......x ....   **(3 marks)**

(b) Put a tick (✔) beside the equation which is the equation of a straight line that is perpendicular to the line with equation $y = 3x - 4$.

$y = 3x - 4$	$y = 4 - 3x$	$y = \frac{1}{3}x - 4$	$y = 4 - \frac{1}{3}x$	$y = 3x + 4$

(1 mark)

3 A straight line **L** passes through the point with coordinates $(3, 7)$ and is perpendicular to the line with equation $y = 3x + 5$. Find the equation of the line **L**.

.... **(3 marks)**

4 $ABCD$ is a rectangle. A is the point $(0, 1)$ and C is the point $(0, 6)$. The equation of the straight line through A and B is $y = 2x + 1$

(a) Find the equation of the straight line through D and C.

.... **(2 marks)**

(b) Find the equation of the straight line through B and C.

.... **(2 marks)**

5 The point P has coordinates $(2, 1)$ and the point Q has coordinates $(-2, -1)$. Find the equation of the perpendicular bisector of PQ.

.... **(4 marks)**

Quadratic graphs

1 (a) Complete the table of values for $y = x^2 - 2$

> Guided

x	-3	-2	-1	0	1	2	3
y		2					7

> Substitute each value of x into the equation $y = x^2 - 2$ to find the value of y.

$x = -3$: $y = (-3 \times -3) - 2 = \ldots\ldots - 2 = \ldots\ldots\ldots\ldots$

$x = -1$: $y = (-1 \times \ldots\ldots) - 2 = \ldots\ldots - 2 = \ldots\ldots\ldots\ldots$ **(2 marks)**

(b) On the grid draw the graph of $y = x^2 - 2$

(1 mark)

(c) Write down the coordinates of the turning point.

> The turning point is the point where the direction of the curve changes.

.... **(2 marks)**

2 (a) Complete the table of values for $y = x^2 - 4x + 3$

> Guided

x	-1	0	1	2	3	4	5
y		3					8

$x = -1$: $y = (-1 \times -1) - (4 \times -1) + 3 = \ldots\ldots\ldots\ldots$

$x = 1$: $y = (1 \times \ldots\ldots) - (4 \times \ldots\ldots) + 3 = \ldots\ldots\ldots\ldots$ **(2 marks)**

(b) On the grid draw the graph of $y = x^2 - 4x + 3$

(2 marks)

(c) Write down the coordinates of the turning point.

.... **(1 mark)**

Cubic and reciprocal graphs

1 (a) Complete the table of values for $y = x^3 - 4x - 2$

x	-2	-1	0	1	2	3
y			-2			13

> Substitute each value for x into the rule $y = x^3 - 4x - 2$ to find the value of y.

$x = -2$: $y = (-2 \times -2 \times -2) - (4 \times -2) - 2 = $ ....

$x = -1$: $y = (-1 \times -1 \times -1) - (4 \times -1) - 2 = $ ....   **(2 marks)**

(b) On the grid draw the graph of $y = x^3 - 4x - 2$

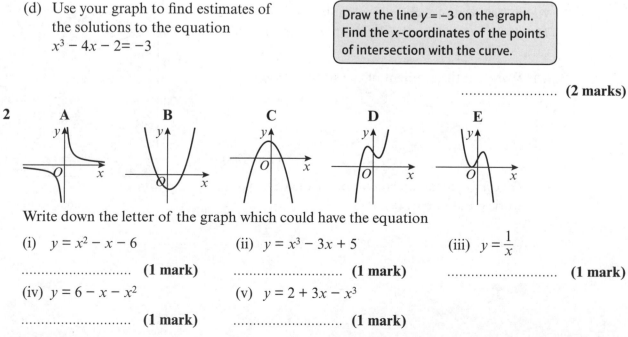

(2 marks)

(c) Write down the coordinates of the turning points.

.... **(2 marks)**

(d) Use your graph to find estimates of the solutions to the equation $x^3 - 4x - 2 = -3$

> Draw the line $y = -3$ on the graph. Find the x-coordinates of the points of intersection with the curve.

.... **(2 marks)**

2

A	B	C	D	E

Write down the letter of the graph which could have the equation

(i) $y = x^2 - x - 6$ (ii) $y = x^3 - 3x + 5$ (iii) $y = \dfrac{1}{x}$

.... **(1 mark)**   .... **(1 mark)**   .... **(1 mark)**

(iv) $y = 6 - x - x^2$ (v) $y = 2 + 3x - x^3$

.... **(1 mark)**   .... **(1 mark)**

Real-life graphs

1 You can use this graph to convert between miles and kilometres.

(a) Use the graph to change 60 miles into kilometres.

Guided

60 miles = ... kilometres

> Draw a vertical line from 60 miles up to the line and then draw a horizontal line across to the kilometres axis.

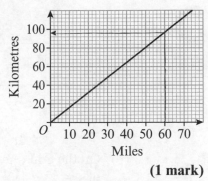

(1 mark)

(b) The distance from Rome to Lyon is 660 miles. The distance from Rome to Marseille is 950 km. Is Rome closer to Lyon or Marseille? You must show all of your working.

> Convert 60 miles to km. Using this information work out 600 miles and then add on the conversion for 60 miles.

(3 marks)

2 Here are four flasks. Rachael fills each flask with water. The graphs show the rate of change of the depth of the water in each flask as Rachael fills it. Draw a line from each flask to the correct graph. One line has been drawn for you.

Guided

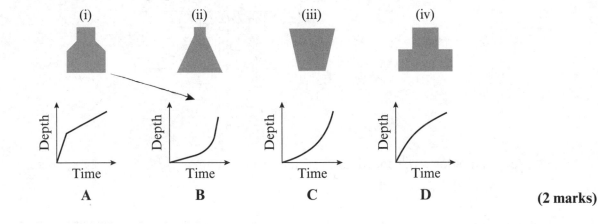

(2 marks)

3 Water flows out of a cylindrical tank at a constant rate. The graph shows how the depth of water in the tank varies with time.

(a) Work out the gradient of the straight line.

.... **(2 marks)**

(b) Give a practical interpretation of the value you worked out in part (a).

.... **(1 mark)**

Quadratic equations

1 Solve

Guided

(a) $x^2 - 3x = 0$

$x(x - ......) = O$

$x = O$ or $x = ...$

(2 marks)

(b) $2x^2 + 5x = 0$

....

(2 marks)

> Find the values of x that make each factor equal to 0. The first factor is just x so one solution Is $x = 0$.

2 Solve

Guided

(a) $x^2 - 4 = 0$

$(x + ...)(x - ......) = O$

$x = ...$ or $x = ...$

(2 marks)

(b) $16x^2 - 49 = 0$

....

(2 marks)

(c) $2x^2 - 50 = 0$

....

(2 marks)

3 Solve

Guided

(a) $x^2 + 6x + 8 = 0$

$(x + 2)(x + ...) = O$

$x = -2$ or $x = ...$

(2 marks)

(b) $x^2 - 7x + 12 = 0$

....

(2 marks)

(c) $x^2 = 10x - 21$

....

(2 marks)

4 Solve

(a) $2x^2 + 7x + 3 = 0$

$(2x + 1)(x + ...) = O$

$x = ...$ or $x = ...$

(2 marks)

> The first factor is $2x + 1$, so the first solution is $x = -\frac{1}{2}$.

(b) $3x^2 + 5x + 2 = 0$

.... **(2 marks)**

5 A rectangle has length $(x + 5)$ cm and width $(x - 2)$ cm. The area of the rectangle is 60 cm^2.

(a) Show that $x^2 + 3x - 70 = 0$.

> Use information given to form an equation for the area. Then rearrange the equation to the form given.

(2 marks)

(b) Solve the equation to find the length and width of the rectangle.

(3 marks)

6 The rectangle shown has length $(x + 16)$ cm and width x cm. The square has length $(3x - 2)$ cm. The area of the rectangle is 16 cm^2 more than the area of the square. The area of the square is more than 1 cm^2. Find the value of x.

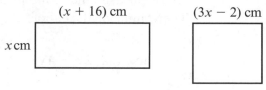

$(x + 16)$ cm $(3x - 2)$ cm

x cm

.... **(4 marks)**

The quadratic formula

1 How many solutions do the following quadratics have?

> Use $b^2 - 4ac$ to determine the number of solutions.

(a) $x^2 - 2x + 1 = 0$

(b) $11x^2 + 2x - 7 = 0$

(c) $3x^2 + 5x + 9 = 0$

.... **(2 marks)**   .... **(2 marks)**   .... **(1 mark)**

2 Solve these quadratic equations and give your answers correct to 3 significant figures.

(a) $x^2 - 5x + 2 = 0$

> Use the quadratic formula
> $$x = \frac{-b \pm \sqrt{b^2 - 4ac}}{2a}$$

$a = 1, b = -5, c = 2$

$$x = \frac{-(-5) \pm \sqrt{(-5)^2 - (4 \times 1 \times 2)}}{2 \times 1}$$

$$= \frac{\dots + \sqrt{\dots}}{\dots} \text{ or } \frac{\dots - \sqrt{\dots}}{\dots} = \dots \text{ or } \dots$$

$$= \dots \text{ or } \dots \text{ (to 3 s.f.)}$$ **(3 marks)**

(b) $2x^2 + 3x - 1 = 0$

(c) $5x^2 - 4x - 2 = 0$

(d) $2 - 3x - x^2 = 0$

.... **(3 marks)**   .... **(3 marks)**   .... **(3 marks)**

3 The diagram shows a six-sided shape.
All the corners are right angles.
All the measurements are given in cm.
The area of the shape is 95 cm^2.

(a) Show that $2x^2 + 6x - 95 = 0$.

x

$2x + 1$

5

x

(3 marks)

(b) Work out the value of x. Give your answer correct to 3 significant figures.

$x = \dots$ **(4 marks)**

4 The diagram shows a cuboid.
All the measurements are in cm.
The volume of the cuboid is 51 cm^3.

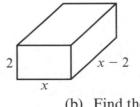

2

$x - 2$

x

(a) Show that $2x^2 - 4x - 51 = 0$.

(b) Find the value of x.

(3 marks) $x = \dots$ **(3 marks)**

(c) What assumption have you made about the value of x?

.... **(1 mark)**

Completing the square

1 Write the following in the form $(x + p)^2 + q$

   (a) $x^2 - 10x + 1$ (b) $x^2 - 2x + 7$ (c) $x^2 + 6x - 4$

> **Guided**

$x^2 - 10x + 1$

$= (x - \text{..........})^2 - \text{..........} + 1$

$= (x - \text{..........})^2 - \text{..........}$ ....

          (2 marks)   **(2 marks)**   **(2 marks)**

2 Write the following in the form $a(x + p)^2 + q$

   (a) $2x^2 + 12x + 9$ (b) $3x^2 + 6x - 2$ (c) $4x^2 + 8x - 11$

> **Guided**

$2x^2 + 12x + 9$

$= 2(x^2 + 6x) + 9$

$= 2(x + \text{..........})^2 - \text{..........} + 9$

$= 2(x + \text{..........})^2 - \text{..........}$ ....

          (2 marks)   **(2 marks)**   **(2 marks)**

3 Solve, giving your answers to 3 significant figures

> Rewrite the equation in completed square form, and then solve. Remember that any positive number has two square roots: one positive and one negative.

   (a) $x^2 + 4x - 6 = 0$ (b) $x^2 - 7x + 3 = 0$ (c) $4x^2 + 7x - 2 = 0$

.... **(2 marks)**   .... **(2 marks)**   .... **(2 marks)**

4 Solve, giving your answers in surd form

   (a) $x^2 + 4x - 3 = 0$ (b) $2x^2 + 5x - 6 = 0$

.... **(2 marks)**   .... **(2 marks)**

5 (a) Find the values of p and q such that $3x^2 + 5x + 1 = a(x + p)^2 + q$

$p = \text{.....................}$

$q = \text{.....................}$ **(2 marks)**

   (b) Hence, or otherwise, solve the equation $3x^2 + 5x + 1 = 0$. Give your answers in surd form.

.... **(2 marks)**

Simultaneous equations 1

Guided

1 Solve the simultaneous equations

(a) $2x + 5y = 16$ ①
    $5x - 2y = 11$ ②

① × 5 gives

......x + ...y = ...   ③

② × 2 gives

$10x - 4y = 22$ ④

③ − ④ gives

......y = ...

   y = ...

Substitute y = ... in ①:

$2x + 5$ ...... = 16

        x = ...

        x = ..., y = ... **(3 marks)**

> Label the equations ① and ②.

(b) $3x + 2y = 11$
    $2x - 5y = 20$

x = ..., y = ... **(3 marks)**

2 By drawing two suitable straight lines on the coordinate grids below, solve the simultaneous equations

Guided

(a) $x + y = 5$
    $y = 3x + 1$

(b) $2x + y = 5$
    $x + y = 3$

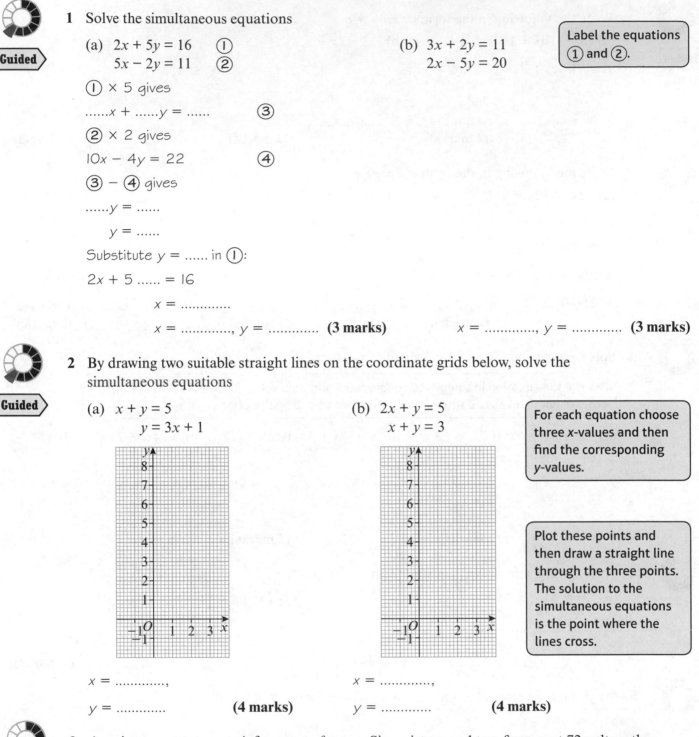

> For each equation choose three x-values and then find the corresponding y-values.

> Plot these points and then draw a straight line through the three points. The solution to the simultaneous equations is the point where the lines cross.

x = ...,

y = ... **(4 marks)**

x = ...,

y = ... **(4 marks)**

3 A resistor costs r pence. A fuse costs f pence. Six resistors and two fuses cost 72p altogether.

(a) Show that $3r + f = 36$

(1 mark)

Five resistors and three fuses cost 80p altogether.

(b) Work out the cost of one resistor and the cost of one fuse.

Resistor = ... p

Fuse = ... p   **(4 marks)**

Simultaneous equations 2

Guided

1 Solve the simultaneous equations

$$y = x + 1$$
$$x^2 + y^2 = 5$$

$$x^2 + y^2 = 5$$

$$x^2 + (\ldots + \ldots)^2 = 5$$

$$x^2 + x^2 + \ldots x + \ldots = 5$$

$$2x^2 + \ldots x - \ldots = 0$$

$$x^2 + \ldots x - \ldots = 0$$

$$(x + \ldots)(x - \ldots) = 0$$

$$x = \ldots \text{ and } x = \ldots$$

$$y = \ldots + 1 = \ldots \text{ and } y = \ldots + 1 = \ldots$$

> Always substitute the linear equation into the quadratic equation.

> Multiply out the brackets and then solve the quadratic equation.

> Cancel.

> Factorise the expression.

> Substitute these values into the linear equation.

(5 marks)

2 Solve the simultaneous equations

$$y = 6 - 2x$$
$$xy + x = 3$$

.... **(5 marks)**

3 Solve the simultaneous equations

$$x + y = 4$$
$$y = 2x^2 - 7x + 8$$

> Rearrange $x + y = 4$ to $y = \ldots$

.... **(5 marks)**

4 L is the straight line with equation $y = x - 3$
C is the circle with equation $x^2 + y^2 = 45$
The line intersects the circle at two points.
Find the coordinates of both points.

.... and .... **(5 marks)**

Equation of a circle

Guided

1   Write down the equation of a circle with

   (a) centre $(0, 0)$ and radius 6.

   (b) centre $(0, 0)$ and radius 13.

   (c) centre $(0, 0)$ and radius $\frac{5}{4}$.

$x^2 + y^2 = r^2$

$x^2 + y^2 = $ .... **(1 mark)**   .... **(1 mark)**   .... **(1 mark)**

2   (a) Solve the simultaneous equations

$$y = x - 2$$
$$x^2 + y^2 = 2$$

.... **(5 marks)**

   (b) Use your answer to part (a) to state the geometrical relationship between the line $y = x - 2$ and the circle $x^2 + y^2 = 2$

.... **(1 mark)**

3   (a) Draw the graph of $x^2 + y^2 = 25$.

                                              (2 marks)

   (b) Draw the graph of $y = x + 2$ on the same axes.

                                              (1 mark)

   (c) Use your graph to write down the solutions to the simultaneous equations $y = x + 2$ and $x^2 + y^2 = 25$.

.... **(2 marks)**

4   **C** is the circle with equation $x^2 + y^2 = 52$

   (a) Verify that the point $(4, -6)$ lies on **C**.

PROBLEM SOLVED!

You will need to use problem-solving skills throughout your exam – **be prepared!**

(1 mark)

   (b) **T** is a tangent to the circle at the point $(4, -6)$. Find the equation of **T**.

.... **(5 marks)**

Inequalities

1 Solve

(a) $2x - 3 \leqslant 5$ $(+ 3)$

$2x \leqslant 5 +$ ....

$2x \leqslant$ .... $(\div 2)$

$x \leqslant$ ....   **(2 marks)**

(b) $6x - 3 > 2x + 9$

.... **(2 marks)**

(c) $3 + x < 25 + 3x$

.... **(2 marks)**

2 Find the integer values of x that satisfy both of these inequalities.

(a) $x - 5 > 3$ and $3x + 1 < 31$

$x - 5 > 3$ $(+ 5)$

$x > 3 +$ ....

$x >$ .... **(3 marks)**

(b) $2x + 1 > 7$ and $4x - 3 < 17$

.... **(3 marks)**

3 Find the integer values of x that satisfy both of these inequalities.

$3(x - 2) > x - 4$ and $4x + 12 < 2x + 18$

.... **(3 marks)**

4 Solve the inequalities

(a) $\dfrac{8x - 5}{3} \geqslant 9$

.... **(3 marks)**

(b) $\dfrac{2x + 3}{3} < \dfrac{3x + 5}{7}$

.... **(3 marks)**

5 A rectangular patio has length x m. The length is 7 m more than the width.
The perimeter of the patio must be less than 53 m.

(a) Form a linear inequality in x.

.... **(2 marks)**

(b) Solve your inequality to find the greatest value of x where x is an integer.

$x =$ .... **(1 mark)**

Quadratic inequalities

Guided

1 Solve these inequalities. Represent your solutions on the number line.

(a) $x^2 - 9 > 0$

$(x + \text{.......})(x - \text{.......}) > 0$

```
  |———————————○————————————○———————————
 −6 −5 −4 −3 −2 −1  0  1  2  3  4  5
```

.... **(2 marks)**

(b) $x^2 - 25 \geqslant 0$

> Factorise the quadratic.

```
 |—————————————————————————————————————
−6 −5 −4 −3 −2 −1  0  1  2  3  4  5  6
```

.... **(2 marks)**

(c) $x^2 \leqslant 16$ $x^2 - 16 \leqslant 0$

```
 |—————————————————————————————————————
−6 −5 −4 −3 −2 −1  0  1  2  3  4  5  6
```

.... **(2 marks)**

(d) $x^2 > 1$

```
 |—————————————————————————————————————
−6 −5 −4 −3 −2 −1  0  1  2  3  4  5  6
```

.... **(2 marks)**

2 Solve these inequalities. Leave your answers in surd form.

(a) $x^2 - 27 < 0$

.... **(2 marks)**

(b) $x^2 - 48 \geqslant 0$

.... **(2 marks)**

3 Solve these inequalities.

(a) $x^2 - 4x - 5 < 0$

.... **(2 marks)**

> You are interested in the values of x where the graph is below the horizontal axis.

> Start by sketching the graph of $y = x^2 - 4x - 5$.

(b) $x^2 - 11x + 24 \geqslant 0$

.... **(2 marks)**

(c) $x^2 - 3x - 10 \leqslant 0$

.... **(2 marks)**

(d) $4x^2 - 8x + 3 > 0$

.... **(2 marks)**

4 (a) Sketch the graph of $y = 3 - 2x - x^2$

.... **(2 marks)**

(b) Hence, or otherwise, solve the inequality $3 - 2x - x^2 < 0$

.... **(2 marks)**

Trigonometric graphs

1 (a) Sketch the graph of $y = \tan x$ in the range $0 \leqslant x \leqslant 360°$. **(2 marks)**

(b) State the equation of an asymptote.

.... **(2 marks)**

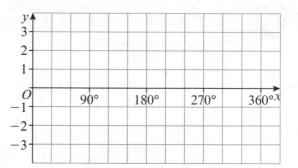

2 Find the two values of angle A, $0 \leqslant A \leqslant 360°$, such that

(a) $\sin A = 0.5$

> Find the first angle.

$A = \sin^{-1} 0.5$

$A = ...°$

Second angle = ...° − ....° = ... **(2 marks)**

(b) $\cos A = \dfrac{\sqrt{3}}{2}$ (c) $\tan A = 0.68$

.... **(2 marks)**   .... **(2 marks)**

3 The diagram shows the graph of $y = \cos x$ for $0 \leqslant x \leqslant 360°$

(a) Mark on the diagram the two solutions of $\cos x = -0.5$
    (1 mark)

(b) Write down the two solutions of $\cos x = -0.5$

.... **(2 marks)**

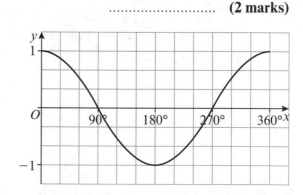

(c) State the coordinates of the turning points.

.... **(2 marks)**

4 (a) On the axes, sketch the graph of $y = \sin x$ for $0 \leqslant x \leqslant 360°$. Label the axes and mark the scales clearly.
    (3 marks)

(b) Hence, or otherwise, state the two values of angle x, $0 \leqslant x \leqslant 360°$, such that $\sin x = \dfrac{\sqrt{3}}{2}$

.... **(2 marks)**

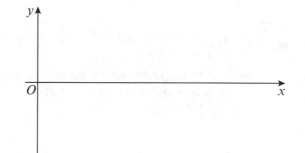

(c) State the coordinates of the turning points.

.... **(1 mark)**

Transforming graphs

1 The graph of $y = f(x)$ is shown on each diagram. Sketch the graph of

(a) $y = f(x) + 2$

(b) $y = -f(x)$

(2 marks)   **(2 marks)**

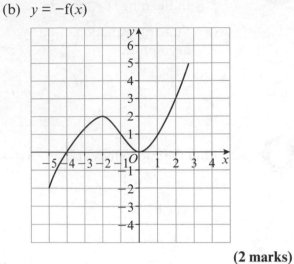

2 The diagram shows part of the curve with equation $y = f(x)$

The coordinates of the maximum point of this curve are (2, 3).

Write down the coordinates of the maximum point of the curve with equation

(a) $y = f(x - 4)$

(b) $y = f(x) + 4$

.... **(1 mark)**   .... **(1 mark)**

(c) $y = -f(x)$

(d) $y = f(-x)$

.... **(1 mark)**   .... **(1 mark)**

3 The curve with equation $y = x^2$ is translated so that the point at (0, 0) is mapped onto the point (4, 0). Find an equation of the translated curve.

> **Guided**

$y = (x\,..........................)^{....}$

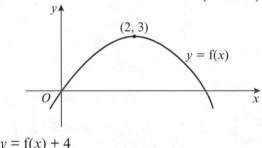

(2 marks)

4 This graph shows the curve $y = \sin x$ for $0 \le x \le 360°$

On the same axes sketch the graph of $y = -\sin x$ for $0 \le x \le 360°$.

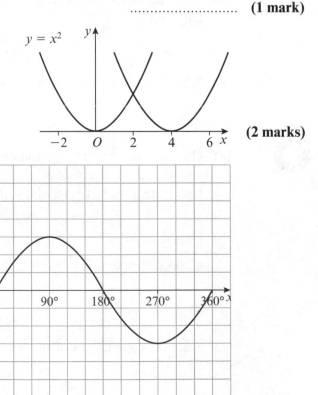

(2 marks)

Inequalities on graphs

1 (a) Shade in the region which satisfies these inequalities

$3y + 2x \geqslant 12$ $y < x - 1$ $x < 6$

(2 marks)

Draw the graph of $x = 6$.

Use dotted lines (for < and >) and solid lines (for ⩽ and ⩾). Points on a solid line are included in the region but points on a dotted line aren't.

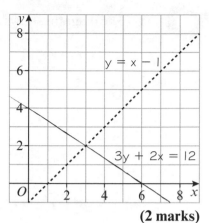

(b) x and y are integers. On the grid, mark with a cross (×) each of the **four** points which satisfies **all** three inequalities.

(2 marks)

2 The line with equation $6y + 5x = 15$ is drawn on the grid.

(a) (i) On the grid, shade the region of points whose coordinates satisfy these four inequalities.

$y \geqslant 0$ $x > 0$ $2x < 3$ $6y + 5x \leqslant 15$

(ii) Label this region **R**.   **(2 marks)**

P is a point in the region **R**.
The coordinates of P are both integers.

(b) Write down the coordinates of P.

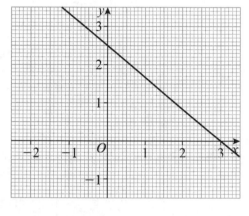

.... **(1 mark)**

3 (a) On the grid below, draw straight lines and use shading to show the region **R** that satisfies these inequalities.

$x > 2$ $y \leqslant x$ $x + y \leqslant 6$ $y > 0$

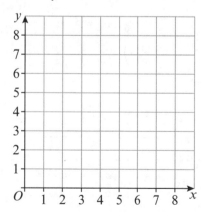

(3 marks)

(b) The point P with coordinates (x, y) lies inside the region **R**. x and y are integers. Write down the coordinates of all the possible points of P.

.... **(2 marks)**

Using quadratic graphs

1 The diagram shows a graph of $y = x^2 - 3x + 1$

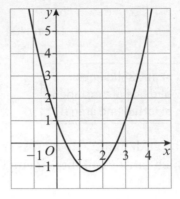

(a) Find estimates for the solutions of $x^2 - 3x + 1 = 0$

.... **(1 mark)**

(b) Use a graphical method to find estimates for the solutions to the equation
$x^2 - 3x + 1 = x - 1$

.... **(2 marks)**

2 (a) Complete the table of values for $y = x^2 - 4x$

x	-1	0	1	2	3	4	5
y		0	-3		-3		5

(2 marks)

(b) On the grid, draw the graph of $y = x^2 - 4x$

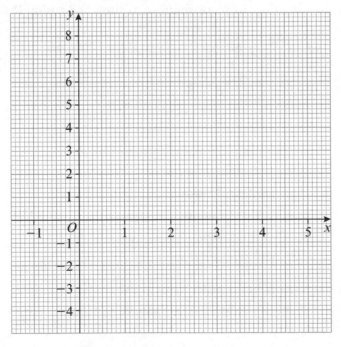

(2 marks)

(c) Find estimates for the solutions of $x^2 - 4x = 0$

.... **(2 marks)**

(d) By drawing a suitable straight line, work out the solutions of the equation $x^2 - 5x + 2 = 0$

.... **(2 marks)**

Turning points

1 Write down the coordinates of the turning points of the following curves.

(a) $y = x^2 + 4x - 6$

$y = (x + \text{.......})^2 - \text{.......} - 6 = (x + \text{.......})^2 - \text{.......}$

Turning point = (......., ...) **(3 marks)**

(b) $y = x^2 + 2x - 4$ (c) $y = x^2 + 3x + 2$ (d) $y = 2x^2 + 3x - 7$

.... **(3 marks)**   .... **(3 marks)**   .... **(3 marks)**

2 (a) By writing the quadratic equation $y = x^2 - 4x + 2$ in the form $(x + p)^2 + q$, find p and q.

$p = \text{.....................}$

$q = \text{.....................}$ **(3 marks)**

(b) Hence, or otherwise, write down the minimum value of the equation.

.... **(1 mark)**

PROBLEM SOLVED!

3 Sketch the following graphs of $y = f(x)$, showing the coordinates of the turning point and the coordinates of any intercepts with the coordinate axes.

> You will need to use problem-solving skills throughout your exam – **be prepared!**

(a) $f(x) = x^2 + 3x + 10$ (b) $f(x) = x^2 - 4x + 4$

(4 marks)   **(4 marks)**

4 The diagram shows a graph of the curve with equation $y = f(x)$

(a) Write down the coordinates of the turning point.

.... **(1 mark)**

The curve with equation $y = f(x)$ is a translation of the curve with equation $y = x^2$

(b) Find the equation of the curve $y = f(x)$.

.... **(2 marks)**

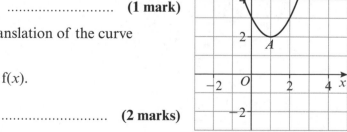

43

Sketching graphs

1 Look at the graphs sketched below. Write down any intercepts with the coordinate axes.

> **Guided**

(a) $y = x^2 + 4x - 21$

(b) $y = 3 - 5x - 2x^2$

$y = (x + \text{.......})(x - \text{.......})$

$x = \text{.......}$ and $x = \text{.......}$ **(2 marks)**

.... **(2 marks)**

2 Look at the graphs sketched below. Write down any intercepts with the coordinate axes.

> **Guided**

(a) $y = x(x - 3)(x + 3)$

(b) $y = (x - 4)(x - 2)(x + 5)$

> If one factor of a cubic equation is x, then $x = 0$ is one solution.

$x(x - 3)(x + 3) = 0$

$x = \text{.......}, x = \text{.......}$ and $x = \text{.......}$ **(2 marks)**

.... **(2 marks)**

(c) $y = (x - 1)^2(x + 2)$

(d) $y = (x + 4)^2(x - 1)$

.... **(2 marks)**

.... **(2 marks)**

3 Sketch the following graphs of $y = f(x)$, showing the coordinates of any intercepts with the coordinate axes.

(a) $y = x^2 - 8x + 15$

(b) $y = x(x^2 + 3x - 18)$

(2 marks)

(3 marks)

Iteration

1 (a) Show that when $f(x) = 0$, the equation $f(x) = x^3 + x - 3$ can be rearranged to give $x = \sqrt[3]{3 - x}$

Guided

$$x^3 + x - 3 = 0$$

$$x^3 = \text{.......} - \text{.......}$$

$$x = \sqrt[3]{\text{.......} - \text{.......}}$$

> Rearrange the equation for x^3

> Take the cube root of the right-hand side.

(2 marks)

(b) Use the iterative formula $x_{n+1} = \sqrt[3]{3 - x_n}$ with $x_0 = 1$ to find the real root of $f(x)$ correct to 2 decimal places.

> Work out x_1, x_2, x_3 and so on. Write down all the digits shown on your calculator display, then round each value to 2 d.p. until you reach two values that both round to the same number.

.... **(3 marks)**

2 (a) Show that when $f(x) = 0$, the equation $f(x) = x^3 - 7x + 2$ can be rearranged to give $x = \sqrt[3]{7x - 2}$

.... **(1 mark)**

(b) Use the iterative formula $x_{n+1} = \sqrt[3]{7x_n - 2}$ with $x_0 = -2.4$ to find the real root of $f(x)$ correct to 3 decimal places.

.... **(3 marks)**

3 (a) Complete the table of values for $y = x^3 - 5x - 8$

x	-3	-2	-1	0	1	2	3
y	-20			-8	-12		4

(2 marks)

(b) On the grid, draw the graph of $y = x^3 - 5x - 8$ **(2 marks)**

(c) Show that when $f(x) = 0$, the equation $f(x) = x^3 - 5x - 8$ can be rearranged to give $x = \sqrt[3]{5x + 8}$

(2 marks)

(d) Use the iterative formula $x_{n+1} = \sqrt[3]{5x_n + 8}$ to find the real root of $f(x)$ correct to 3 decimal places.

.... **(3 marks)**

Rearranging formulae

1 Rearrange $y = \frac{1}{3}x + 4$ to make x the subject.

$$y = \frac{1}{3}x + 4 \qquad (-4)$$

$$y - \text{.......} = \frac{1}{3}x \qquad (\times 3)$$

$$\text{.......}(y - \text{.......}) = x$$

> Insert brackets around the expression on the left-hand side.

(2 marks)

2 Make h the subject of the formula $d = \sqrt{\frac{3h}{2}}$

$$d = \sqrt{\frac{3h}{2}} \qquad \text{(square both sides)}$$

$$d^{\cdots} = \frac{3h}{2} \qquad (\times 2)$$

$$\text{.......}d^{\cdots} = 3h \qquad (\div 3)$$

$$h = \text{.......} d^{\cdots}$$

(2 marks)

3 Rearrange the formula $P = \frac{3w + 10}{120}$ to make w the subject.

.... **(2 marks)**

4 Rearrange the formula $P = \pi r + 2r + 2x$ to make x the subject.

.... **(2 marks)**

5 $M = \frac{t^2 + x}{t + x}$

Rearrange the formula to make x the subject.

$$M = \frac{t^2 + x}{t + x} \qquad [\times (t + x)]$$

$$M(t + x) = t^2 + x \qquad \text{(expand brackets)}$$

$$M\text{.......} + M\text{.......} = t^2 + x \quad (-t^2 \text{ and } -Mx)$$

$$M\text{.......} - \text{.......} = x - \text{.......}$$

$$M\text{.......} - \text{.......} = x(\text{.......} - \text{.......})$$

$$\frac{M\text{.......} - \text{.......}}{\text{.......} - \text{.......}} = x$$

> Multiply both sides by the denominator.

> Collect all the x terms on one side.

> Factorise the right-hand side.

(3 marks)

6 $y = \frac{2pt}{p - t}$

Rearrange the formula to make t the subject.

.... **(3 marks)**

Algebraic fractions

1 Simplify fully

  (a) $\dfrac{3}{4x} + \dfrac{1}{8x}$

> Find the common denominator.

$\dfrac{\ldots\ldots}{\ldots\ldots x} + \dfrac{\ldots\ldots}{\ldots\ldots x} = \dfrac{\ldots\ldots + \ldots\ldots}{\ldots\ldots x} = \dfrac{\ldots\ldots}{\ldots\ldots x}$

(2 marks)

  (c) $\dfrac{2}{x+3} - \dfrac{1}{x-2}$

.... **(2 marks)**

  (b) $\dfrac{x+1}{2} - \dfrac{x-2}{3}$

.... **(2 marks)**

  (d) $\dfrac{x}{y} + \dfrac{y}{x}$

.... **(2 marks)**

2 Simplify fully

  (a) $\dfrac{x^2 + 2x + 1}{4x + 4}$

> Factorise the numerator and the denominator and then cancel.

$\dfrac{(x + \ldots\ldots)(x + \ldots\ldots)}{\ldots\ldots (x + \ldots\ldots)} = \dfrac{(x + \ldots\ldots)}{\ldots\ldots}$

(2 marks)

  (b) $\dfrac{x^2 + 4x - 5}{x^2 + 2x - 3}$

.... **(2 marks)**

3 Simplify fully

  (a) $\dfrac{3}{x} \times \dfrac{x}{5}$

> Cancel first.

$\dfrac{3}{\cancel{x}} \times \dfrac{\cancel{x}}{5} = \ldots\ldots\ldots$ **(2 marks)**

  (c) $\dfrac{5}{xy} \div \dfrac{x}{y}$

.... **(2 marks)**

  (b) $\dfrac{2r^2}{5} \times \dfrac{4}{r^3}$

.... **(2 marks)**

  (d) $\dfrac{x+2}{x+4} \div \dfrac{3x+6}{x^2-16}$

.... **(2 marks)**

4 Solve

  (a) $\dfrac{3x}{4} + \dfrac{2}{3} = x$

$x = $ .... **(3 marks)**

  (c) $\dfrac{3x+2}{2} - \dfrac{x-1}{5} = 3$

$x = $ .... **(3 marks)**

  (b) $\dfrac{3}{4x} - \dfrac{1}{2x} = 4$

$x = $ .... **(3 marks)**

  (d) $\dfrac{5}{x+1} - \dfrac{3}{2(x+1)} = \dfrac{1}{2}$

.... **(3 marks)**

Quadratics and fractions

Guided

1 Solve

(a) $\dfrac{3}{x} + \dfrac{1}{x-4} = 1$

> Multiply everything by $x(x-4)$ to remove the fractions.

$$\dfrac{3......(...... - ......)}{x} + \dfrac{1......(...... - ......)}{x-4} = 1......(...... - ......)$$

> Cancel.

$$3(...... - ......) + 1...... = x(...... - ......)$$

> Multiply out and simplify.

$$......x - ...... + 1x = x^{....} - ......$$

> Rearrange as a quadratic equation.

$$x^{....} - ......x + ... = 0$$

> Solve the quadratic equation using factorisation.

$$(x......)(x......) = 0$$

$$x = ... \text{ and } x = ...$$

(4 marks)

(b) $\dfrac{2}{x-3} + \dfrac{1}{x-4} = 2$

.... **(4 marks)**

(c) $\dfrac{2}{x+1} - \dfrac{3}{2x+3} = 1$

.... **(4 marks)**

(d) $\dfrac{3}{2x-1} - \dfrac{2}{1-3x} = 4$

.... **(4 marks)**

(e) $\dfrac{1}{x} - \dfrac{2}{x-1} = 6$

.... **(4 marks)**

(f) $\dfrac{x+1}{4} - \dfrac{20}{x-5} = 1$

.... **(4 marks)**

(g) $\dfrac{x+4}{x-2} = x$

.... **(4 marks)**

2 Jack travels a distance of 400 km. He travels at an average speed of x km/h.

(a) Write down an expression in terms of x for the time Jack travels.

.... **(1 mark)**

(b) Jack increases his average speed by 10 km/h for his return journey.
Write down an expression in terms of x for the time taken for his return journey.

.... **(2 marks)**

(c) If Jack's travelling time is 40 minutes less at the faster average speed, show that

$$\dfrac{400}{x} - \dfrac{400}{x+10} = \dfrac{2}{3}$$

.... **(2 marks)**

(d) Work out the value of x. Give your answer correct to 3 significant figures.

.... **(2 marks)**

Surds 2

1 Show that $(4 + \sqrt{3})^2 = 19 + 8\sqrt{3}$

Show each stage of your working clearly.

> When the question says 'Show that …'
> you should start from the left-hand side,
> then simplify and rearrange until your
> expression matches the right-hand side.

Guided

$(4 + \sqrt{3})(4 + \sqrt{3}) = \text{.......} + \text{.......}\sqrt{3} + \text{.......}\sqrt{3} + \text{.......}$

$= \text{.......} + \text{.......}\sqrt{3}$

(3 marks)

2 Show that $(\sqrt{2} + \sqrt{3})^2 = 5 + 2\sqrt{6}$

Show each stage of your working clearly.

(3 marks)

3 Show that $(4 + \sqrt{5})(2 - \sqrt{5}) = 3 - 2\sqrt{5}$

Show each stage of your working clearly.

(3 marks)

4 a and b are positive integers such that $(1 - \sqrt{a})^2 = b - 2\sqrt{7}$

Find the value of a and the value of b.

$a = $ ....

$b = $ .... **(3 marks)**

5 The diagram shows a right-angled triangle. All measurements are in cm.

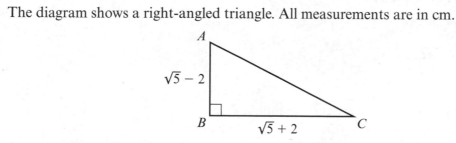

Work out, leaving your answers in surd form where appropriate,

(a) the area of the triangle

.... cm^2 **(3 marks)**

(b) the length of AC.

.... cm **(3 marks)**

Functions

f and g are functions such that $f(x) = 2x - 1$ and $g(x) = \frac{2}{x}$, $x \neq 0$

(a) Find the value of

> **Guided**

>> Substitute $x = 7$ into $f(x)$ and simplify.

  (i) $f(7)$

  $f(7) = 2 \underline{(7)} - 1 = \underline{13}$ **(1 mark)**

  (ii) $fg(3)$

>> Order is important. Substitute $x = 3$ into $g(x)$ first and then substitute this answer into $f(x)$

$\frac{2}{3}$ $f(\frac{2}{3})$ $2(\frac{2}{3}) - 1 = \frac{1}{3}$ **(1 mark)**

(b) Find $gf(x)$

>> Substitute $2x - 1$ into $g(x)$

$g(x) = \dfrac{2}{2x - 1}$ **(2 marks)**

2 f and g are functions. $f(x) = 4x - 3$ $g(x) = 1 + x^3$

  (a) Find $f(-2)$

  (b) Given that $g(a) = 28$, find the value of a.

$f(-2) = -11$ **(1 mark)**

$1 + a^3 = 28$
$a = 3$ **(2 marks)**

  (c) Work out $gf(2)$

  (d) Work out $fg(x)$

$f(2) = 4(2) - 3 = 5$
$g(5) = 1 + (5)^3$
$= 126$ $\underline{126}$ **(2 marks)**

$f(1+x^3) = 4(1+x^3) - 3$
$= 4x^3 - 3$ **(2 marks)**

3 The function f is defined as $f(x) = \dfrac{x}{x - 2}$, $x \neq 2$

  (a) Find the value of $f(3)$

$\dfrac{3}{3-2} = 3$

3 **(1 mark)**

  (b) Find $ff(x)$. Give your answer in its simplest form.

$\dfrac{(x/x-2)}{(x/x-2)-2}$ $\dfrac{\frac{x}{x-2}}{\frac{x}{x-2}-2}$ **(3 marks)**

4 f and g are functions such that $f(x) = x^2$ and $g(x) = 3x - 2$
  Solve $fg(x) = f(x)$

$f(3x-2) = (3x-2)(3x-2)$

$9x^2 - 6x - 6x + 4$

$9x^2 - 12x + 4 = x^2$

$8x^2 - 12x + 4 = 0$

$x = 1$
$x = \frac{1}{2}$

$1, \frac{1}{2}$ **(4 marks)**

Inverse functions

1 Find the inverse function, f^{-1}, of the function

(a) $f(x) = 3x - 5$

Let $y = 3x - 5$

$\boxed{\text{Make the } x \text{ the subject.}}$

$3x = y + \text{.......}$

$x = \dfrac{y + \text{.......}}{\text{.......}}$

Therefore, $f^{-1} = $ ....   **(2 marks)**

(b) $f(x) = \dfrac{4x}{3} + 7$ (c) $f(x) = \dfrac{x - 7}{x}$

f^{-1} .... **(2 marks)** f^{-1} .... **(2 marks)**

2 (a) $f(x) = 3x + 4$ (b) $g(x) = 4x - 1$

　　Find $f^{-1}(x)$ Find $g^{-1}(x)$

$f^{-1}(x)$ .... **(2 marks)** $g^{-1}(x)$ .... **(2 marks)**

(c) Hence, or otherwise, find an expression for $f^{-1}(x) + g^{-1}(x)$
　　You must simplify your answer.

.... **(2 marks)**

3 The function f is such that $f(x) = 2x + 1$

(a) Find $f^{-1}(x)$ (b) State the value of $ff^{-1}(5)$

$f^{-1}(x)$ .... **(2 marks)** $ff^{-1}(x)$ .... **(1 mark)**

4 The function g is such that $g(x) = \dfrac{x}{x - 1}$

(a) Solve the equation $g(x) = \frac{5}{2}$ (b) Find $g^{-1}(x)$

$x = $ .... **(2 marks)** $g^{-1}(x)$ .... **(3 marks)**

Algebraic proof

1 Prove that $(2n - 1)^2 + (2n + 1)^2 = 2(4n^2 + 1)$

 Guided

LHS $= (2n - 1)^2 + (2n + 1)^2$

$= (2n - 1)(2n - 1) + (2n + 1)(2n + 1)$

> Simplify by multiplying out the brackets.

> Add the brackets together and then factorise.

(2 marks)

2 $5(x - c) = 4x - 5$ where c is an integer

Prove that x is a multiple of 5.

> Multiply out the brackets, then rearrange to make x the subject.

(3 marks)

3 Prove that $(3x + 1)^2 - (3x - 1)^2$ is a multiple of 4, for all positive integer values of x.

(3 marks)

4 Prove that the sum of any three consecutive even numbers is always a multiple of 6.

PROBLEM SOLVED!

> You will need to use problem-solving skills throughout your exam – **be prepared!**

> Write your even numbers as $2n$, $2n + 2$ and $2n + 4$

(3 marks)

5 A rectangular number is defined as $n(n + 1)$. Prove that the sum of two consecutive rectangular numbers is always double a square number.

(3 marks)

6 The nth term of a sequence is given by $x_n = \dfrac{1}{n}$

Prove that $x_n - x_{n+1} = \dfrac{1}{n(n + 1)}$

(3 marks)

7 Prove that the difference between the square of any 2-digit number and the square of that number when reversed has a common factor of 99.

> For example:
> $73^2 - 37^2 = 3960$
> $ = 99 \times 40$

(5 marks)

Exponential graphs

1 Sketch the graph of

(a) $y = 3^x$ (b) $y = 5^x$ (c) $y = 2^{-x}$

(2 marks)   **(2 marks)**   **(2 marks)**

Guided

2 The sketch shows a curve with equation $y = ka^x$
   where k and a are constants, and $a > 0$.
   Show that the value of a is 4 times the value of k.

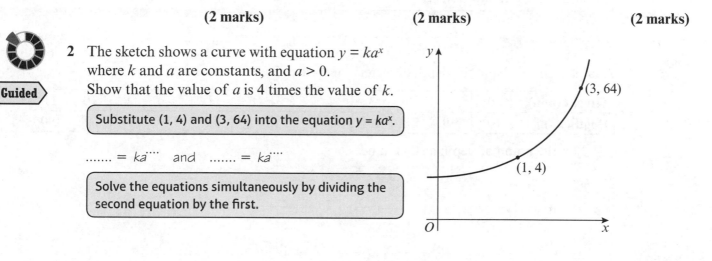

> Substitute (1, 4) and (3, 64) into the equation $y = ka^x$.

.... $= ka^{....}$ and   .... $= ka^{....}$

> Solve the equations simultaneously by dividing the
> second equation by the first.

(4 marks)

3 A scientist is investigating the growth in population of a certain bacteria. At 12 noon,
   there are 2000 bacteria in a Petri dish. The population of bacteria grows exponentially
   at the rate of 20% per hour.

(a) Show that the population after 3 hours is 3456.

(2 marks)

(b) The number of bacteria, N, in the Petri dish h hours after 12 noon can be
     modelled by the formula $N = a \times b^h$ where a and b are to be determined.
     Write down the value of a and the value of b.

$a =$ ....

$b =$ .... **(2 marks)**

The population of bacteria in the Petri dish after 12 hours is k times the population
of the bacteria after 9 hours.

(c) Find the value of k.

$k =$ .... **(2 marks)**

Gradients of curves

Guided

1 A fish tank is filled with water. The graph shows how the diameter of the surface of the water changes with time.

Estimate the gradient when the time is 10 seconds.

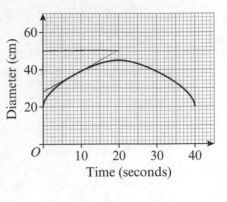

> The gradient of the curve at any point is the gradient of the tangent at that point. Draw a tangent at $t = 10$. Complete the triangle and work out the horizontal distance and the vertical distance.

gradient of tangent $= \dfrac{\text{vertical distance}}{\text{horizontal distance}} = \dfrac{\text{.......} - \text{.......}}{\text{.......} - \text{.......}} = $ .... **(2 marks)**

2 A tank is emptied and the depth of water is recorded. Here are the results.

Time, t (min)	0	1	2	3	4	5	6	7	8	9	10	11
Depth, d (m)	6.0	5.9	5.8	5.6	5.4	5.2	4.9	4.5	3.9	3.0	1.8	0.0

(a) Plot the graph of depth against time.

(2 marks)

(b) Work out the average rate of decrease of depth between $t = 0$ and $t = 11$.

.... m/min **(2 marks)**

(c) Estimate the average rate of decrease of depth when the depth is 5 m.

.... m/min **(2 marks)**

3 In an experiment, Michaela heated a liquid to 64 °C then allowed it to cool. The graph shows her results.

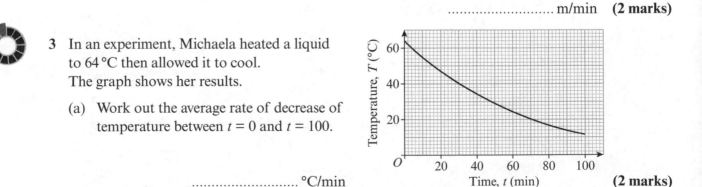

(a) Work out the average rate of decrease of temperature between $t = 0$ and $t = 100$.

.... °C/min **(2 marks)**

(b) Estimate the average rate of decrease of temperature at $t = 40$.

.... °C/min **(2 marks)**

Velocity–time graphs

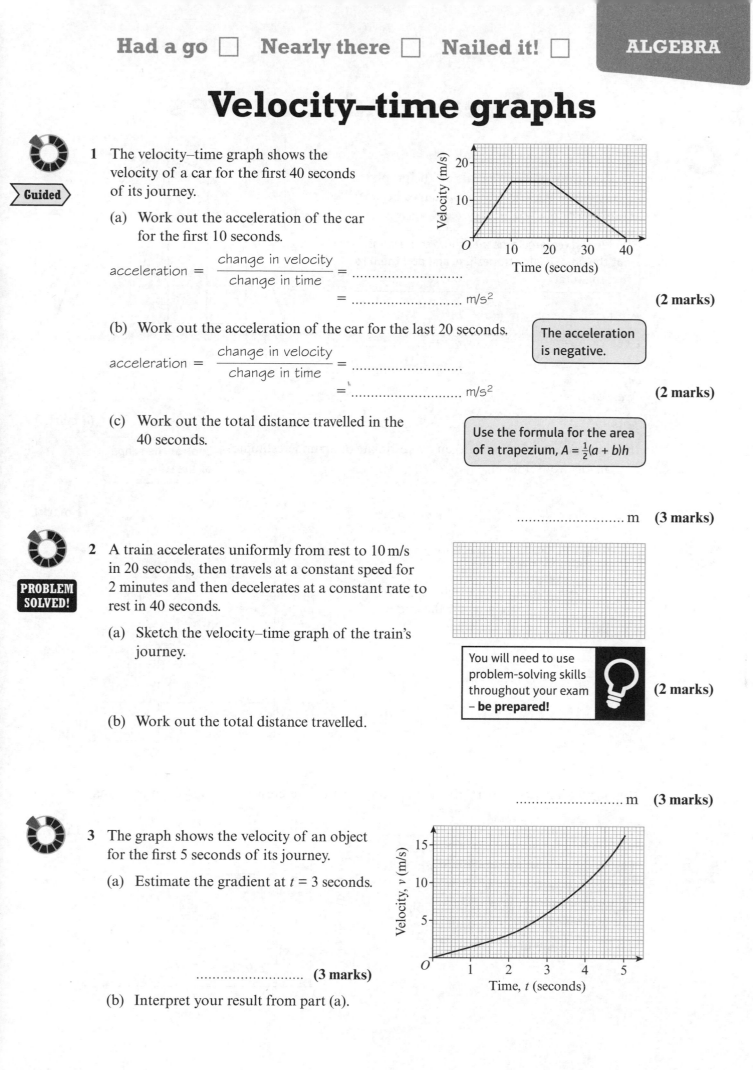

1 The velocity–time graph shows the velocity of a car for the first 40 seconds of its journey.

> Guided

   (a) Work out the acceleration of the car for the first 10 seconds.

   $$\text{acceleration} = \frac{\text{change in velocity}}{\text{change in time}} = \text{..........................}$$

   $$= \text{..........................} \text{ m/s}^2$$ **(2 marks)**

   (b) Work out the acceleration of the car for the last 20 seconds.

   > The acceleration is negative.

   $$\text{acceleration} = \frac{\text{change in velocity}}{\text{change in time}} = \text{..........................}$$

   $$= \text{..........................} \text{ m/s}^2$$ **(2 marks)**

   (c) Work out the total distance travelled in the 40 seconds.

   > Use the formula for the area of a trapezium, $A = \frac{1}{2}(a + b)h$

   .... m   **(3 marks)**

2 A train accelerates uniformly from rest to 10 m/s in 20 seconds, then travels at a constant speed for 2 minutes and then decelerates at a constant rate to rest in 40 seconds.

> **PROBLEM SOLVED!**

   (a) Sketch the velocity–time graph of the train's journey.

   > You will need to use problem-solving skills throughout your exam – **be prepared!**   **(2 marks)**

   (b) Work out the total distance travelled.

   .... m   **(3 marks)**

3 The graph shows the velocity of an object for the first 5 seconds of its journey.

   (a) Estimate the gradient at $t = 3$ seconds.

   .... **(3 marks)**

   (b) Interpret your result from part (a).

   .... **(2 marks)**

Areas under curves

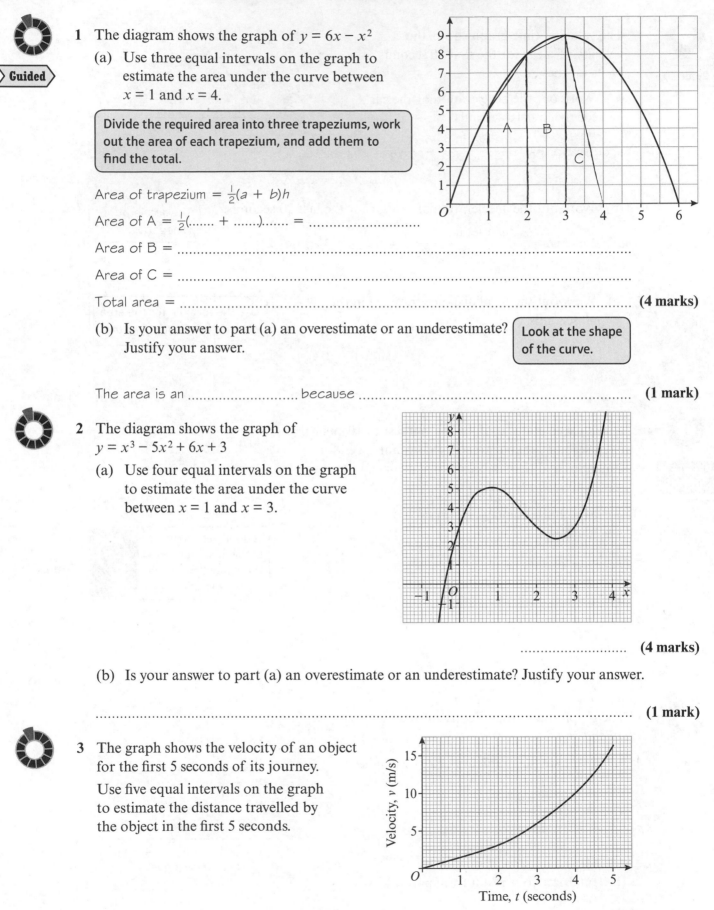

1 The diagram shows the graph of $y = 6x - x^2$

(a) Use three equal intervals on the graph to estimate the area under the curve between $x = 1$ and $x = 4$.

> Divide the required area into three trapeziums, work out the area of each trapezium, and add them to find the total.

Area of trapezium = $\frac{1}{2}(a + b)h$

Area of A = $\frac{1}{2}$(....... + ...)....... = ...

Area of B = ...

Area of C = ...

Total area = ... **(4 marks)**

(b) Is your answer to part (a) an overestimate or an underestimate? Justify your answer.

> Look at the shape of the curve.

The area is an .... because .... **(1 mark)**

2 The diagram shows the graph of $y = x^3 - 5x^2 + 6x + 3$

(a) Use four equal intervals on the graph to estimate the area under the curve between $x = 1$ and $x = 3$.

.... **(4 marks)**

(b) Is your answer to part (a) an overestimate or an underestimate? Justify your answer.

.... **(1 mark)**

3 The graph shows the velocity of an object for the first 5 seconds of its journey.

Use five equal intervals on the graph to estimate the distance travelled by the object in the first 5 seconds.

.... m **(4 marks)**

Problem-solving practice 1

1 The diagram shows a trapezium $ABCD$ with AD parallel to BC.

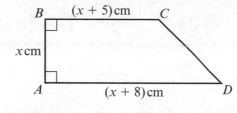

$AB = x$ cm $BC = (x + 5)$ cm $AD = (x + 8)$ cm

The area of the trapezium is 42 cm^2.

Show that x is a square number.

(4 marks)

2 x and y are positive integers such that $(3 - \sqrt{x})^2 = y - 6\sqrt{5}$

Work out the value of x and the value of y.

$x = $ ....

$y = $ .... **(3 marks)**

3 The diagram shows two straight lines, AB and CD. The coordinates of points A, B and C are $A(1, 3)$, $B(5, 9)$ and $C(0, 8)$. Point D is on AB and is halfway between point A and point B.

Marcie says that line CD is not perpendicular to line AB. Is she correct?

You must show all your working.

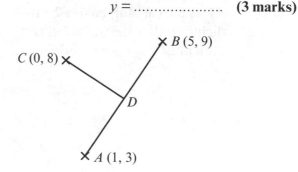

(4 marks)

4 Here are the first five terms of a quadratic series.

1   1   3   7   13

Work out an expression, in terms of n, for the nth term of this sequence.

.... **(4 marks)**

Problem-solving practice 2

5 The diagram shows two rods, *AB* and *BC*, joined together.

A ├──── $\left(\dfrac{x+3}{4}\right)$ m ────┤ B ├── $\left(\dfrac{x-5}{3}\right)$ m ──┤ C

Rod *AB* is $\left(\dfrac{x+3}{4}\right)$ m in length and rod *BC* is $\left(\dfrac{x-5}{3}\right)$ m.

The total length of the rods is 9 m.

Work out the difference, in m, between the lengths of rod *AB* and rod *BC*.

.... m   **(4 marks)**

6 A car approaches a set of traffic lights.
The traffic lights turn red.
The driver brakes and the car slows
down to a stop. The distance, in metres,
from the time the driver brakes to the
traffic lights is 40 m.
Here is the speed–time graph for the
5 seconds until the car stops.
Does the car stop before the traffic
lights or after the traffic lights?
You must show your working.

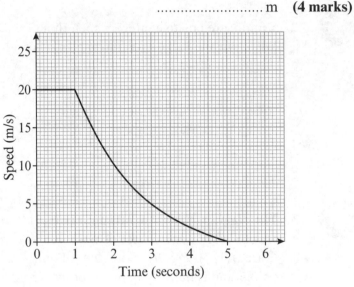

(4 marks)

7 Find the coordinates of the points where line $y + 5 = 3x$ intersects the
circle $x^2 + y^2 = 65$

.... **(5 marks)**

Calculator skills 2

1 In a sale, normal prices are reduced by 12%. The normal price of a television was £720. Work out the amount of the reduction in price of the television.

> Find 12% of £720

Guided

12 ÷ 100 = ...

Reduction = ... × £720 = £...........................

(3 marks)

2 Between 2005 and 2015 the population of Indonesia increased from 224 500 000 to 255 700 000.

(a) Work out the population increase from 2005 and 2015. ....

(b) Write your answer to part (a) as a percentage of 224 500 000 ....

(3 marks)

3 In May 2013 the population of the USA was 310 million. 40 million of these American people spoke Spanish as their first language.

(a) Express 40 million as a percentage of 310 million. Give your answer correct to 1 decimal place.

...........................% **(2 marks)**

(b) Between May 2013 and May 2015 the population of the USA increased by 6%. Work out 6% of 310 million. Give your answer correct to the nearest million.

.... **(3 marks)**

4 Ravina is going on holiday to France. The exchange rate is £1 = €1.420 13. Ravina changes £475 into euros.

(a) Work out how many euros she should receive. Give your answer to the nearest euro.

€.......................... **(2 marks)**

(b) Ravina wants to buy a blouse. The cost of the blouse is €144. She uses an approximation of £2 = €3. Using this approximation, work out an approximate cost of the blouse in pounds.

£.......................... **(3 marks)**

(c) Is it reasonable to use the approximation £2 = €3 to work out the cost of the blouse in pounds instead of using the exact exchange rate £1 = €1.420 13? You must give a reason for your answer.

.... **(1 mark)**

Ratio

1 (a) Divide £50 in the ratio 2 : 3.

Total parts = ... + ...

   1 part = 50 ÷ .... = ...

2 parts = ... × .... = ...

3 parts = ... × .... = ...   **(2 marks)**

(b) Divide £750 in the ratio 2 : 5 : 8.

.... **(2 marks)**

2 Amish, Benji and Cary save some money in the ratio 3 : 4 : 9.
Cary saved £120 more than Benji.

(a) Show that Amish saved £72.

(3 marks)

(b) Show that the total amount of money saved was £384.

(2 marks)

3 Solder is made from lead and tin.
The ratio of the weight of lead to the weight of tin is 2 : 3.
Kyle made 70 grams of solder. Work out the weight of the lead used.

Total parts = ... + ...

   1 part = 70 ÷ .... = ...

2 parts = 2 × .... = ...   **(2 marks)**

4 Gabby and Harry shared some money based on their ages. The ratio of Gabby's age to Harry's age is 3 : 8 Harry received £2000 more than Gabby.
How much money did they share?

You will need to use problem-solving skills throughout your exam – **be prepared!**

PROBLEM SOLVED!

   8 − 3 = 5 parts

5 parts = £2000

| Harry received £2000 more than Gabby and 8 − 3 = 5 so 5 parts of the ratio represents £2000. |

   1 part = £2000 ÷ 5 = ...

11 parts = 11 × .... = ...

In total they shared ....   **(3 marks)**

Proportion

Guided

1 10 men take 8 days to build a wall.
How long will it take 4 men to do
the same job?

> Work out how long it will take
> 1 person to build the wall.

10 men work 8 days

> Will 4 men take more time (×)
> or less time (÷)?

I man works .... × .... = ...

4 men work .... ÷ .... = ...

.... days   **(2 marks)**

2 A large basket of sweets costs £4.80 and holds 200 g. A medium basket of sweets costs
£4.50 and holds 175 g. Which size basket is better value for money?

> Show all your working and
> then write a conclusion.

(3 marks)

3 A school building can be decorated by 12 men working 8 hours a day for 5 days.
Mike wants to know how long it would take 10 men working 6 hours a day.

.... days   **(2 marks)**

4 Here is a list of the ingredients needed to make scones for 4 people. Gemma wants
to make scones. for 10 people. Does she have enough ingredients? You must show all
your working.

Scones for 4
200 g flour
2 eggs
50 g currants
100 ml milk

Gemma's ingredients
0.7 kg flour
5 eggs
240 g currants
0.2 litres milk

> Work out how much
> of each ingredient is
> needed for 10 scones

(4 marks)

5 2.25 kg of a type of ham costs €28.32 in Germany.
$2\frac{3}{4}$ lb of the same ham costs £12.42 in Scotland.
In which country is it cheaper to buy the ham? Show all of your working.
1 kg = 2.2 lb   £1 = €1.39

(3 marks)

Percentage change

Guided

1 Aaron is comparing the cost of flights from two airlines.
Both airlines charge a credit card charge and a booking fee.
- Mega-jet charges 3% for using a credit card and a £3.50 booking fee.
- Air-whizz charges 5% for using a credit card and a £2.10 booking fee.

A ticket is advertised as costing £90 from both airlines.
Work out which airline is cheaper after the additional charges are applied.

Mega-jet

$\frac{3}{....}$ × .... = £.........................

£......................... + £......................... + £......................... = £.........................

Air-whizz

$\frac{5}{....}$ × .... = £.........................

£......................... + £......................... + £......................... = £.........................

.... is cheaper.   **(4 marks)**

2 Noah and Chloe are collecting reward points in an online video game.

(a) Noah collected 3200 points last month and 4315 points this month.
Work out the percentage increase in the number of points he collected.

...........................% **(3 marks)**

(b) Chloe collected 5100 points last month and 3672 points this month.
Work out the percentage decrease in the number of points she collected.

...........................% **(3 marks)**

3 Niamh and Owen work for the same company.
In 2014 Niamh earned £24 500 per year. In 2015 she received a pay rise to £25 970.
In 2014 Owen earned £22 000. In 2015 he received the same percentage pay rise as Niamh.
Work out Owen's salary in 2015.

£......................... **(4 marks)**

PROBLEM SOLVED!

4 The price of a holiday in 2013 was £1450. The same holiday cost an extra 14% in 2014. In 2015 the same holiday was reduced by 12% of its price in 2014. Work out the price of this holiday in 2015.

> You will need to use problem-solving skills throughout your exam – **be prepared!**

£......................... **(4 marks)**

5 In 2014, Brickworld made 150 000 clay bricks and a profit of £2080. In 2015, it increased its production by 18%. It costs £45 to produce 1000 bricks. Brickworld sells a crate of 100 bricks for £6. Assuming all the bricks are sold, work out the percentage profit for 2015.

...........................% **(4 marks)**

Reverse percentages

1 In a sale all prices are reduced by 30%. Andy buys a shirt on sale for £42.
Work out the original price of the shirt.

Guided

100% − 30% = ...%

$\frac{......}{100}$ = ...

£42 ÷ .... = £..............

> First work out the multiplier for a 30% decrease.

(3 marks)

2 Brinder receives a pay rise of 6%. After the pay rise Brinder earns a salary of £35 245.
Work out Brinder's salary before the pay rise.

Guided

100% + ...% = ... %

$\frac{......}{100}$ = ...

£35 245 ÷ .... = £..............

> First work out the multiplier for a 6% increase.

(3 marks)

3 Kam bought a new car. The car depreciates by 15% each year. After one year the car was worth £28 560. Work out the price of the car when it was new.

> Check that your answer makes sense. The original price of the car should be greater than £28 560.

£.......................... **(3 marks)**

4 Between 2014 and 2015 a large company increased its workforce by 4%. Following this increase it had 780 employees. Work out the numbers of employees at the company in 2014.

.... **(3 marks)**

5 Kate's weekly wage this year is £560. This is 8% more than her weekly wage from last year. Ken says, 'Your weekly wage was £515.20 last year'. Is Ken correct? You must show your working.

> You can do this question without using reverse percentages. Increase £515.20 by 8% and compare your answer to £560. Remember to write a conclusion.

(3 marks)

6 Alison and Nav invested some money in the stock market in 2014. This table shows the value of their investments in 2015.

PROBLEM SOLVED!

> You will need to use problem-solving skills throughout your exam – **be prepared!**

	Value in 2015	Percentage increase since original investment
Alison	£1848	12%
Nav	£1764	5%

Who invested more money originally? Give reasons for your answer.

(4 marks)

Growth and decay

1 Raj invests £12 000 for 4 years at 10% per annum compound interest. Work out the value of the investment at the end of 4 years.

> **First work out the multiplier for a 10% increase.**

Guided

100% + 10% = ...%

$\frac{......}{100}$ = ...

£12 000 × .... = £............... **(2 marks)**

2 Neil invests £5800 at a compound interest rate of 6% per annum. At the end of *n* complete years the investment has grown to £6907.89. Work out the value of *n*.

> **Choose some values of *n* and work out the amount of investment after *n* years.**

.... **(2 marks)**

3 Chris bought a lorry that had a value of £24 000. Each year the value of the lorry depreciates by 15%.

Guided

(a) Work out the value of the lorry at the end of four years.

100% − ....% = ...%

> **First work out the multiplier.**

$\frac{......}{100}$ = ...

£24 000 × ....$^{......}$ = £............... **(2 marks)**

(b) Shelley bought a new car for £12 000. Each year the value of the car depreciates by 12%. Work out the value of the car at the end of five years.

£.......................... **(2 marks)**

4 Daljit invested £1500 on 1 January 2010 at a compound interest rate of *r*% per annum. The value, £*V*, of this investment after *n* years is given by the formula
$V = 1500 \times (1.065)^n$.

(a) Write down the value of *r*.

r = ... **(1 mark)**

(b) Work out the value of Daljit's investment after 10 years.

£.......................... **(2 marks)**

5 Terry buys a new vacuum cleaner for £350. The value of the machine depreciates by 25% each year. Terry says, '4 × 25% = 100%, so after 4 years the vacuum cleaner will have no value.'
Explain why Terry is wrong.

.... **(2 marks)**

Speed

1   The distance from Manchester to Rome is 1700 km. A plane flies from Manchester to Rome in 4 hours. Work out the average speed of the plane.

This is the formula triangle for speed.

$\dfrac{D}{S \mid T}$

.... km/h   **(2 marks)**

2   David runs 400 metres in 44.7 seconds. Work out his average speed.

Speed – distance ÷ time

Guided

Speed = ... ÷ .... = ... m/s   **(2 marks)**

3   Tracey drives 264 km in 2 hours 45 minutes. Work out Tracy's average speed.

Write 2 hours and 45 minutes in hours.

..............................km/h   **(3 marks)**

4   Chandra travels for 4 hours. Her average speed in her car is 60 km/h. Work out the total distance the car travels.

Guided

Distance = ... × ....

Always write down the formula.

Distance = ... × .... = ...km   **(2 marks)**

5   Pavan is driving in France. The legal speed limit on French motorways is 130 km/h. He travels from one junction to another in 15 minutes and he covers a distance of 35 000 m. Show that he has broken the speed limit.

(3 marks)

6   Jane travelled 50 km in 1 hour 15 minutes. Carol travelled 80 km in 2 hours and 45 minutes. Who had the lower average speed? You must show your working.

(3 marks)

7   At a school's sports day the 100 m race was won in 14.82 seconds and the 200 m was won in 29.78 seconds. Which race was won with a faster average speed? You must show all your working.

(3 marks)

Had a go ☐   Nearly there ☐   Nailed it! ☐

Density

This is the formula triangle for density.

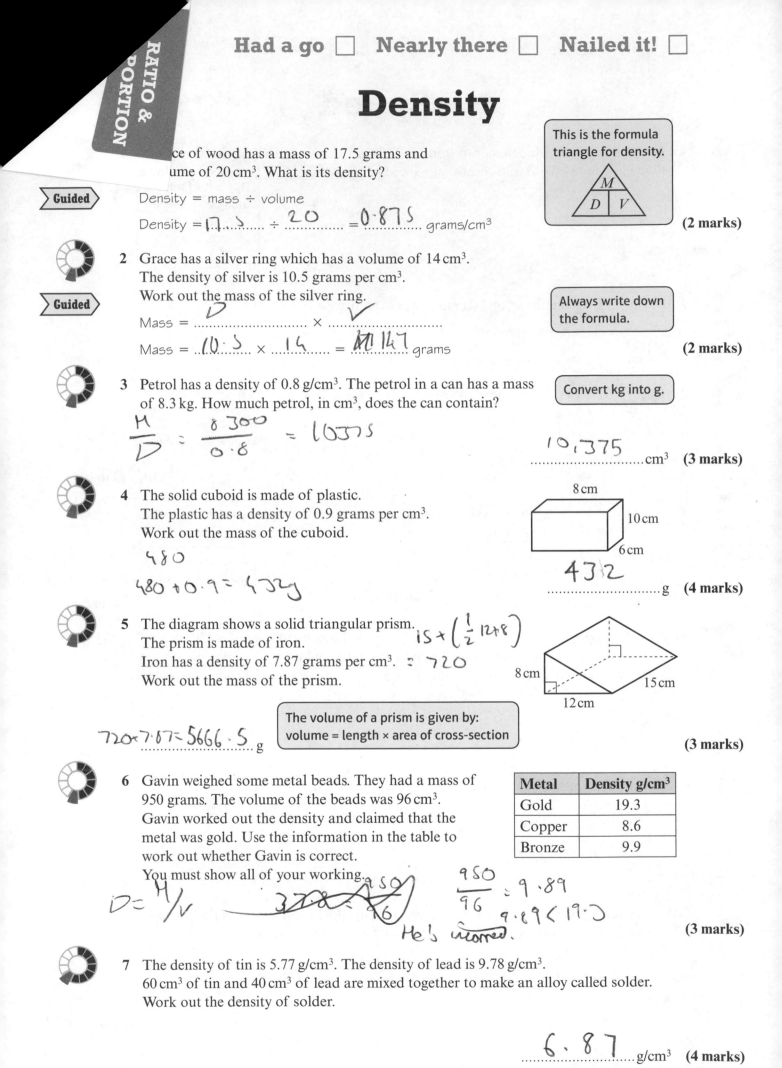

> Guided

...ce of wood has a mass of 17.5 grams and ...ume of 20 cm³. What is its density?

Density = mass ÷ volume

Density = 17.5 ÷ 20 = 0.875 grams/cm³

(2 marks)

2   Grace has a silver ring which has a volume of 14 cm³.
    The density of silver is 10.5 grams per cm³.
    Work out the mass of the silver ring.

> Guided

Always write down the formula.

Mass = D × V

Mass = 10.5 × 14 = 147 grams

(2 marks)

3   Petrol has a density of 0.8 g/cm³. The petrol in a can has a mass of 8.3 kg. How much petrol, in cm³, does the can contain?

Convert kg into g.

$\frac{M}{D} = \frac{8300}{0.8} = 10375$

10,375 cm³   **(3 marks)**

4   The solid cuboid is made of plastic.
    The plastic has a density of 0.9 grams per cm³.
    Work out the mass of the cuboid.

480

480 × 0.9 = 432 g

432

.... g   **(4 marks)**

5   The diagram shows a solid triangular prism.
    The prism is made of iron.
    Iron has a density of 7.87 grams per cm³.
    Work out the mass of the prism.

$15 \times \left(\frac{1}{2} 12 \times 8\right) = 720$

The volume of a prism is given by:
volume = length × area of cross-section

720 × 7.87 = 5666.5 g

(3 marks)

6   Gavin weighed some metal beads. They had a mass of 950 grams. The volume of the beads was 96 cm³.
    Gavin worked out the density and claimed that the metal was gold. Use the information in the table to work out whether Gavin is correct.
    You must show all of your working.

Metal	Density g/cm³
Gold	19.3
Copper	8.6
Bronze	9.9

$D = M/V$ $\frac{950}{96} = 9.89$

9.89 < 19.3

He's incorrect.

(3 marks)

7   The density of tin is 5.77 g/cm³. The density of lead is 9.78 g/cm³.
    60 cm³ of tin and 40 cm³ of lead are mixed together to make an alloy called solder.
    Work out the density of solder.

6.87 g/cm³   **(4 marks)**

Other compound measures

Guided

1 A safe exerts a force of 600 N on to the floor.
The area of the base of the safe is 1.5 m².
Work out the pressure exerted on the floor.

This is the formula triangle for pressure.

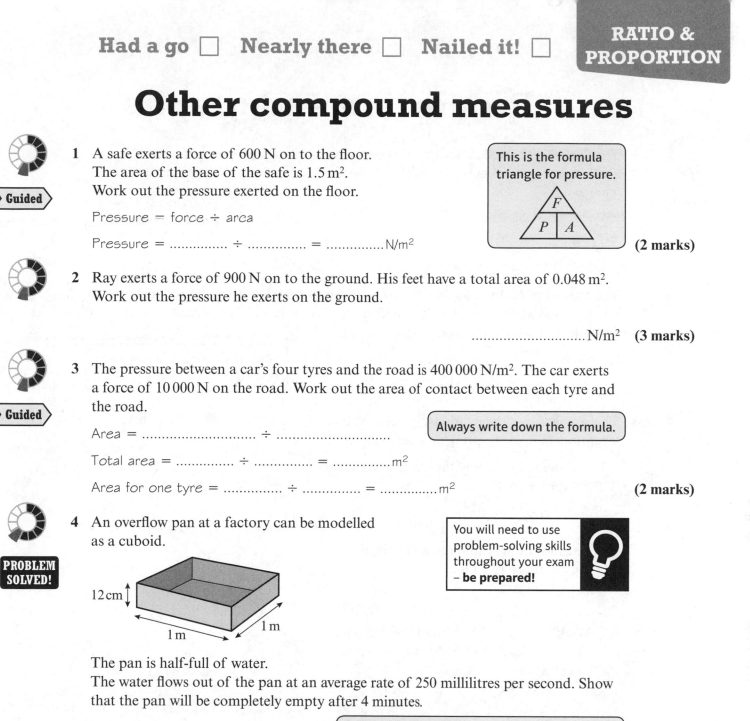

Pressure = force ÷ area

Pressure = ... ÷ .... = ...N/m²

(2 marks)

2 Ray exerts a force of 900 N on to the ground. His feet have a total area of 0.048 m².
Work out the pressure he exerts on the ground.

...........................N/m² **(3 marks)**

Guided

3 The pressure between a car's four tyres and the road is 400 000 N/m². The car exerts a force of 10 000 N on the road. Work out the area of contact between each tyre and the road.

Area = ... ÷ ....

Always write down the formula.

Total area = ... ÷ .... = ...m²

Area for one tyre = ... ÷ .... = ...m²

(2 marks)

PROBLEM SOLVED!

4 An overflow pan at a factory can be modelled as a cuboid.

You will need to use problem-solving skills throughout your exam – **be prepared!**

12 cm

1 m

1 m

The pan is half-full of water.
The water flows out of the pan at an average rate of 250 millilitres per second. Show that the pan will be completely empty after 4 minutes.

1 cm³ = 1 m*l*. Remember to convert metres to cm before calculating the volume of the cuboid.

(3 marks)

5 The area of the Falkland Islands is 12 170 km². The population density is 0.24 people per km². Charles states that the population of the Falkland Islands is 50 708. Is Charles correct? You must show all of your working.

(3 marks)

6 The peak of Ben Nevis can be reached by walking up the Pony Track. The rate of walking is 1200 metres per hour. The Pony Track is 16 km long. Helen begins her journey at 9 am. Can she reach the peak by 11 pm on the same day? You must show all your working.

(3 marks)

Proportion and graphs

1 The force F measured in newtons (N) on a mass is directly proportional to the acceleration a of the mass. Given that $F = 650$ when $a = 25$, work out the value of F when $a = 45$.

> **Guided**

$$\frac{F}{45} = \frac{650}{25}$$

> You can compare ratios to work out F.

$F =$ .... $\times$ .... $=$ .... N **(2 marks)**

2 The resistance, R ohms, of a wire is inversely proportional to the cross-sectional area, A cm^2, of the wire. Given that $R = 30$ when $A = 0.1$, work out the value of R when $A = 0.4$.

> **Guided**

$R \times 0.4 =$ .... $\times$ ....

$R =$ .... $\div$ .... $=$ .... ohms

> When R is inversely proportional to A, you write it as $R \propto \frac{1}{A}$

(2 marks)

3 This graph shows the relationship between the extension, in m, of a spring and the force, in N.

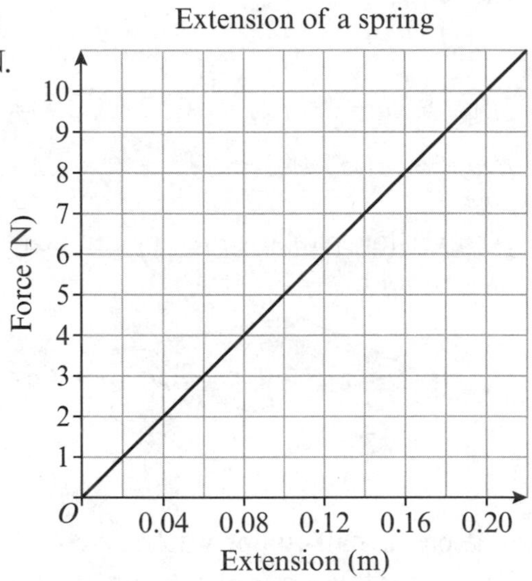

(a) Use the graph to find the extension when the force is 8 N.

.... m **(1 mark)**

(b) Use the graph to find the force when the extension is 0.1 m.

.... n **(1 mark)**

(c) What evidence is there from the graph to show that force is directly proportional to extension?

Extension of a spring

(Graph: y-axis "Force (N)" from 1 to 10, x-axis "Extension (m)" from 0.04 to 0.20, showing a straight line through origin)

.... **(2 marks)**

4 This graph shows the relationship between pressure, in kPa, and volume, in litres.

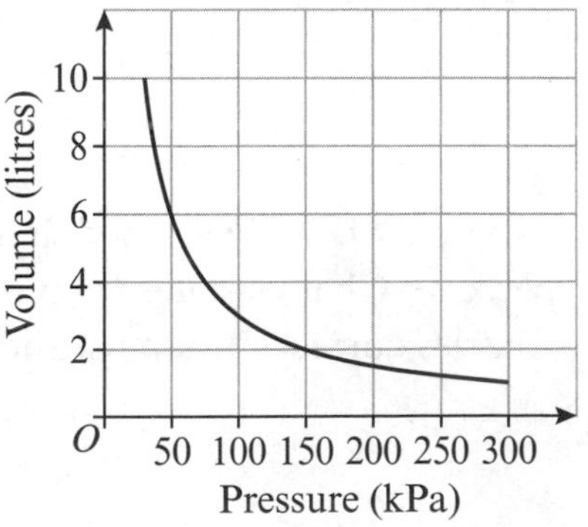

(a) Use the graph to find the volume when the pressure is 150 kPa.

.... litres **(1 mark)**

(b) Use the graph to find the pressure when the volume is 6 litres.

.... kPA **(1 mark)**

(Graph: y-axis "Volume (litres)" from 2 to 10, x-axis "Pressure (kPa)" from 50 to 300, showing a decreasing curve)

(c) What evidence is there from the graph to show that volume is inversely proportional to pressure?

.... **(2 marks)**

Proportionality formulae

1 The time, T seconds, it takes a water heater to boil some water is directly proportional to the mass of water, m kg, in the water heater. When $m = 150$, $T = 900$.

> **Guided**

   (a) Find T when $m = 175$.

   $T \propto m$ $T = km$ $900 = k \times 150$

                                                         $k = \ldots\ldots\ldots\ldots\ldots\ldots\ldots\ldots\ldots$

   $T = \ldots\ldots\ldots\ldots \times \ldots\ldots\ldots\ldots = \ldots\ldots\ldots\ldots$ seconds   **(3 marks)**

The time, T seconds, it takes a water heater to boil a constant mass of water is inversely proportional to the power, P watts, of the water heater. When $P = 1500$, $T = 280$.

   (b) Find the value of T when $P = 700$.

   $T \propto \frac{1}{P}$ $T = \frac{k}{P}$ $k = T \times P$

                                                  $k = \ldots\ldots\ldots\ldots\ldots\ldots \times \ldots\ldots\ldots\ldots\ldots\ldots$

                                                  $k = \ldots\ldots\ldots\ldots\ldots\ldots$

   $T = \dfrac{\ldots\ldots\ldots\ldots}{P}$

   $T = \dfrac{\ldots\ldots\ldots\ldots}{\ldots\ldots\ldots\ldots} = \ldots\ldots\ldots\ldots\ldots$ seconds   **(3 marks)**

2 The weight of a piece of wire is directly proportional to its length. A piece of wire is 100 cm long and has a weight of 24 grams. Another piece of the same wire is 60 cm long. Work out the weight of the 60 cm piece of wire.

                                                              $\ldots\ldots\ldots\ldots\ldots\ldots$ g   **(3 marks)**

3 In a spring, the tension, T newtons, is directly proportional to its extension x cm. When the tension is 176 newtons, the extension is 8 cm.

   (a) Express T in terms of x.

                                                 $T = \ldots\ldots\ldots\ldots\ldots\ldots$ **(3 marks)**

   (b) Calculate the tension, in newtons, when the extension is 14 cm.

                                             $\ldots\ldots\ldots\ldots\ldots\ldots$ N   **(1 mark)**

   (c) Calculate the extension, in cm, when the tension is 319 newtons.

                                             $\ldots\ldots\ldots\ldots\ldots\ldots$ cm   **(2 marks)**

4 f is inversely proportional to d. When $d = 30$, $f = 196$.

   (a) Find the value of f when $d = 70$.   (b) Find the value of d when $f = 24$.

            $\ldots\ldots\ldots\ldots\ldots$ **(3 marks)** $\ldots\ldots\ldots\ldots\ldots$ **(3 marks)**

Harder relationships

1 d is directly proportional to the square of t. $d = 100$ when $t = 5$.

> The square of t can be written as ...².

(a) Express d in terms of t.

Guided

$d \propto t^-$

$d = kt^-$

$100 = k \times 5^-$

$k = $ ....

$d = $ ....t^-

$d = $ .... × .... = ... **(3 marks)**

(b) Work out the value of d when $t = 6$.

$d = $ .... **(1 mark)**

(c) Work out the positive value of t when $d = 64$.

$t = $ .... **(2 marks)**

2 The shutter speed, S, of a camera varies inversely as the square of the aperture setting, f. When $f = 6$, $S = 90$.

(a) Find a formula for S in terms of f.

.... **(3 marks)**

(b) Hence, or otherwise, calculate the value of S when $f = 8$.

$s = $ .... **(1 mark)**

3 The current, I, in an electrical circuit varies as the square root of the power, P. When the current is 4.5 amps the power is 36 watts.

(a) Find a formula for I in terms of P.

.... **(3 marks)**

(b) Work out the value of I, in amps, when $P = 49$ watts.

$I = $ .... **(1 mark)**

(c) Work out the value of P, in watts, when $I = 5$ amps.

$P = $ .... **(2 marks)**

4 T is directly proportional to S^3. When $S = 3$, $T = 324$. Find the value of S when $T = 96$.

$S = $ .... **(4 marks)**

Problem-solving practice 1

1 Karen wants to buy a game for her new PS4. She finds that two online shops sell the game she wants.

> Nile
> Game costs £35.50
> Online discount 16%
> Delivery charge £2.75

> T-bay
> Game costs £30.90 + VAT
> VAT at 20%
> No delivery charge

Karen wants to pay the lowest price. Which shop should Karen buy her game from? You must show all your working.

(4 marks)

2 Angus, Beth and Caitlin save their pocket money. They have saved the money in the ratio 4 : 7 : 12. Caitlin saved £240 more than Angus.

(a) Show that Angus saved £120.

(2 marks)

(b) Work out the total amount of money saved.

£ .... **(2 marks)**

3 Avtar has a full 900 ml bottle of patio sealer.
He is going to mix some of the patio sealer with water.
Here is the information on the label of the bottle.
Avtar is going to use 900 ml of water.
How many millilitres of patio sealer should Avtar use?
You must show your working.

> Patio sealer (900 ml)
> Mix $\frac{1}{5}$ of the patio sealer
> with 5400 ml of water

(4 marks)

Problem-solving practice 2

4   Mario is driving on the motorway in Italy. The speed limit in Italy is 81 km/h.
He drives 325 km in 3 hours 28 minutes. Does Mario break the speed limit? You must
give a reason for your answer.

(3 marks)

5   Brett is going to buy some bird food. Bird food is sold in 200 g boxes costing £2.50 and
in 1000 g boxes costing £10.50. Which box of bird food gives better value for money?
You must show your working.

(3 marks)

6   Kim wants to save £17 500 in 4 years for a deposit on a house. He invests £14 000 in an
ISA for 4 years at 6% per annum compound interest. Will he have enough money after
4 years for a deposit? You must show your working.

(3 marks)

7   The intensity, I candela, of light on an object is inversely proportional to the square of
the distance, x metres, of the object from the light.
The intensity of light is 9 candelas at a distance of 4 metres.

Asha carried out an experiment to test this rule. She works out the intensity of light is
48 candelas when the distance is $\sqrt{3}$ m.
Is she correct? You must show all your working.

(4 marks)

Angle properties

1 Work out the values of x and y.
Give reasons for your answers.

> **Guided**

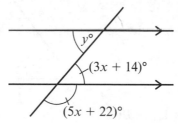

Angles on a straight line add up to ...°.

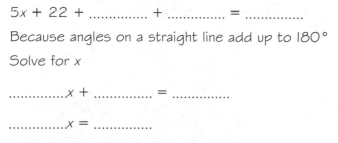

$y°$
$(3x + 14)°$
$(5x + 22)°$

$5x + 22 +$ .... $+$ .... $=$ ....
Because angles on a straight line add up to 180°
Solve for x

...............$x +$ .... $=$ ....

...............$x =$ ....

$x =$ ....

$y = 3x + 14 =$ .... because

.... angles are equal. **(4 marks)**

2 Here is an isosceles triangle.
Work out the values of x and y.

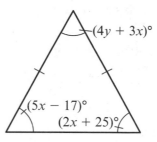

$(4y + 3x)°$

$(5x - 17)°$
$(2x + 25)°$

$x =$ ....

$y =$ .... **(5 marks)**

3 The diagram shows a quadrilateral.
Work out the size of the smallest angle.

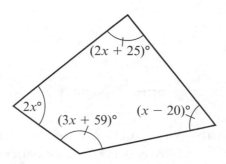

$(2x + 25)°$

$2x°$
$(3x + 59)°$ $(x - 20)°$

.... ° **(4 marks)**

Solving angle problems

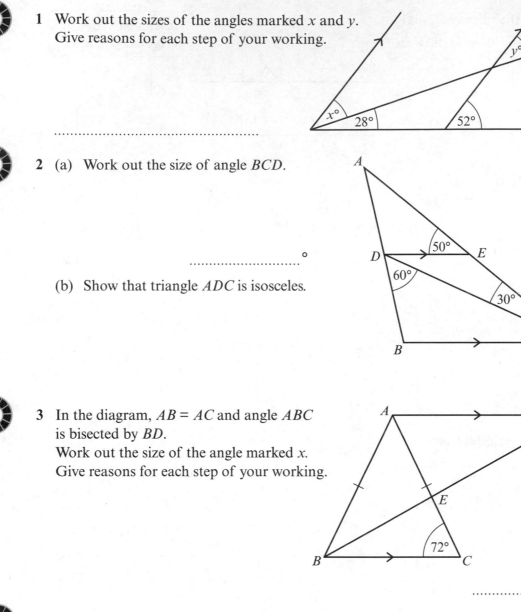

1   Work out the sizes of the angles marked x and y.
    Give reasons for each step of your working.

....   **(4 marks)**

2   (a) Work out the size of angle *BCD*.

.............................°   **(3 marks)**

    (b) Show that triangle *ADC* is isosceles.

                                                                 (2 marks)

3   In the diagram, $AB = AC$ and angle *ABC*
    is bisected by *BD*.
    Work out the size of the angle marked x.
    Give reasons for each step of your working.

.............................° **(4 marks)**

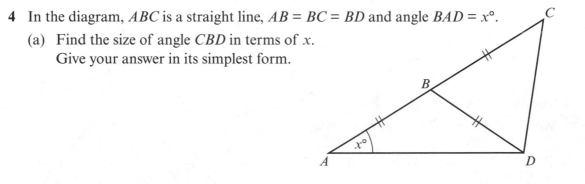

4   In the diagram, *ABC* is a straight line, $AB = BC = BD$ and angle $BAD = x°$.

    (a) Find the size of angle *CBD* in terms of x.
         Give your answer in its simplest form.

.............................° **(2 marks)**

    (b) Work out the size of angle *CDA*.
         Give reasons for each stage of your working.

.............................° **(2 marks)**

Angles in polygons

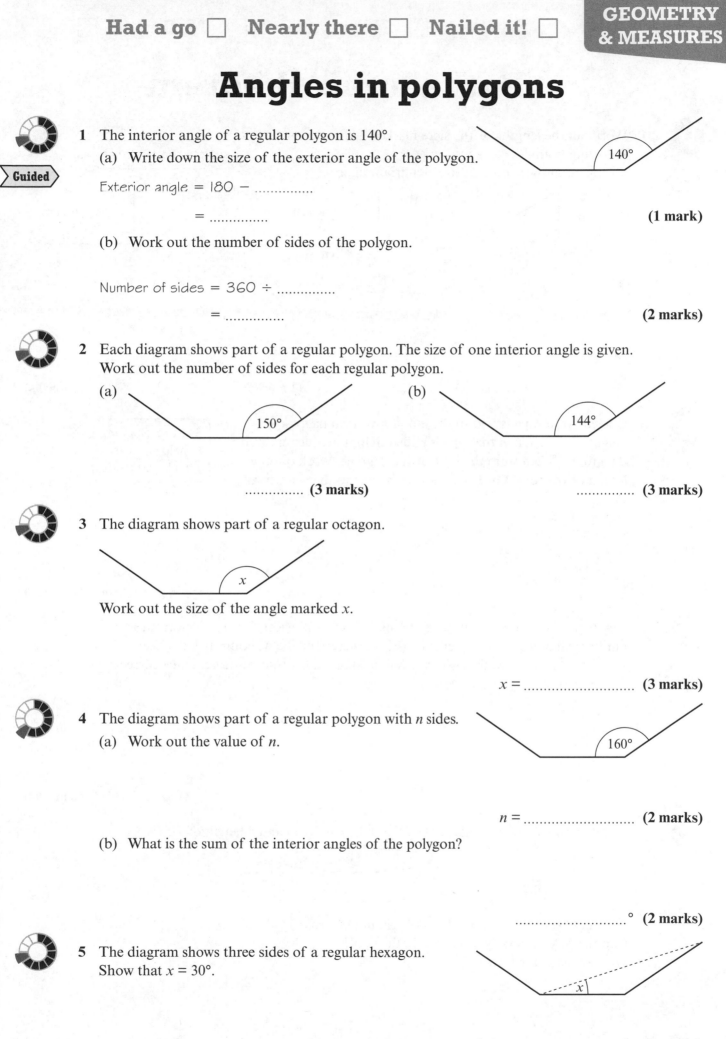

1 The interior angle of a regular polygon is 140°.

  (a) Write down the size of the exterior angle of the polygon.

     Exterior angle = 180 − ....

             = ...   **(1 mark)**

  (b) Work out the number of sides of the polygon.

     Number of sides = 360 ÷ ....

                 = ...   **(2 marks)**

2 Each diagram shows part of a regular polygon. The size of one interior angle is given. Work out the number of sides for each regular polygon.

  (a)   150°   (b)   144°

         .... **(3 marks)**   .... **(3 marks)**

3 The diagram shows part of a regular octagon.

      x

  Work out the size of the angle marked x.

                          x = ... **(3 marks)**

4 The diagram shows part of a regular polygon with n sides.

  (a) Work out the value of n.   160°

                       n = ... **(2 marks)**

  (b) What is the sum of the interior angles of the polygon?

                       ....° **(2 marks)**

5 The diagram shows three sides of a regular hexagon. Show that $x = 30°$.

                  x

                                 (3 marks)

Pythagoras' theorem

1 Work out the lengths of the sides marked with letters in the
following triangles.
Give your answers correct to 3 significant figures.

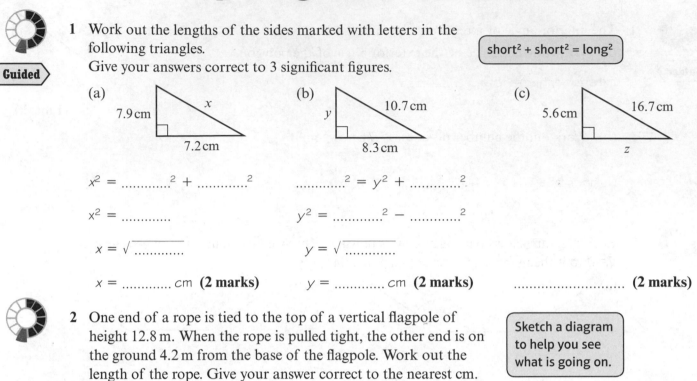

short² + short² = long²

(a)
7.9 cm
x
7.2 cm

(b)
y
10.7 cm
8.3 cm

(c)
5.6 cm
16.7 cm
z

$x^2 = \dots\dots^2 + \dots\dots^2$

$x^2 = \dots\dots$

$x = \sqrt{\dots\dots}$

$x = \dots\dots$ cm   **(2 marks)**

$\dots\dots^2 = y^2 + \dots\dots^2$

$y^2 = \dots\dots^2 - \dots\dots^2$

$y = \sqrt{\dots\dots}$

$y = \dots\dots$ cm   **(2 marks)**

$\dots\dots\dots\dots$ **(2 marks)**

2 One end of a rope is tied to the top of a vertical flagpole of
height 12.8 m. When the rope is pulled tight, the other end is on
the ground 4.2 m from the base of the flagpole. Work out the
length of the rope. Give your answer correct to the nearest cm.

Sketch a diagram
to help you see
what is going on.

$\dots\dots\dots$ cm   **(2 marks)**

3 Cindy has a rectangular suitcase of length 95 cm and width 72 cm. She wants to
put her walking stick into her suitcase. The length of the walking stick is 125 cm.
She thinks that the walking stick will fit into one side of her suitcase. Is she correct?
Give a reason for your answer.

(2 marks)

4 The diagram shows a small pool with a radius of 2.8 m and a height of 1.5 m.

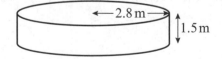

←2.8 m→
1.5 m

A straight pole is 6 m long. The pole cannot be broken.
Can the pole be totally immersed in the pool?
Give a reason for your answer.

(2 marks)

Trigonometry 1

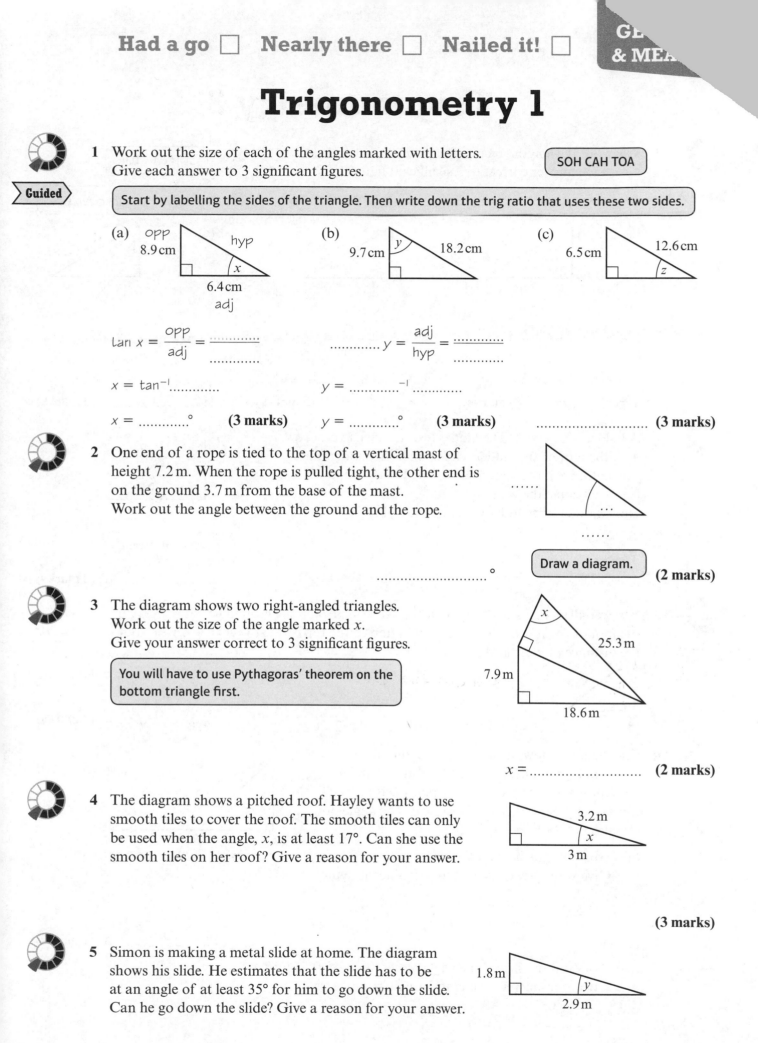

1 Work out the size of each of the angles marked with letters.
Give each answer to 3 significant figures.

SOH CAH TOA

> Guided

Start by labelling the sides of the triangle. Then write down the trig ratio that uses these two sides.

(a) opp
8.9 cm   hyp
        x
6.4 cm
adj

$\tan x = \dfrac{opp}{adj} = \dfrac{............}{............}$

$x = \tan^{-1} ....$

$x = ...°$ **(3 marks)**

(b) 9.7 cm   y   18.2 cm

$............ y = \dfrac{adj}{hyp} = \dfrac{............}{............}$

$y = ...^{-1} ....$

$y = ...°$ **(3 marks)**

(c) 6.5 cm   12.6 cm
            z

.... **(3 marks)**

2 One end of a rope is tied to the top of a vertical mast of
height 7.2 m. When the rope is pulled tight, the other end is
on the ground 3.7 m from the base of the mast.
Work out the angle between the ground and the rope.

Draw a diagram.

............................° **(2 marks)**

3 The diagram shows two right-angled triangles.
Work out the size of the angle marked x.
Give your answer correct to 3 significant figures.

You will have to use Pythagoras' theorem on the bottom triangle first.

x
25.3 m
7.9 m
18.6 m

$x = $ .... **(2 marks)**

4 The diagram shows a pitched roof. Hayley wants to use
smooth tiles to cover the roof. The smooth tiles can only
be used when the angle, x, is at least 17°. Can she use the
smooth tiles on her roof? Give a reason for your answer.

3.2 m
x
3 m

(3 marks)

5 Simon is making a metal slide at home. The diagram
shows his slide. He estimates that the slide has to be
at an angle of at least 35° for him to go down the slide.
Can he go down the slide? Give a reason for your answer.

1.8 m
y
2.9 m

(3 marks)

Trigonometry 2

Work out the length, in cm, of each of the marked sides.
Give each answer correct to 3 significant figures.

SOH CAH TOA

> Guided

Start by labelling the sides of the triangle. Then write down the trig ratio that uses the given and unknown side.

(a)

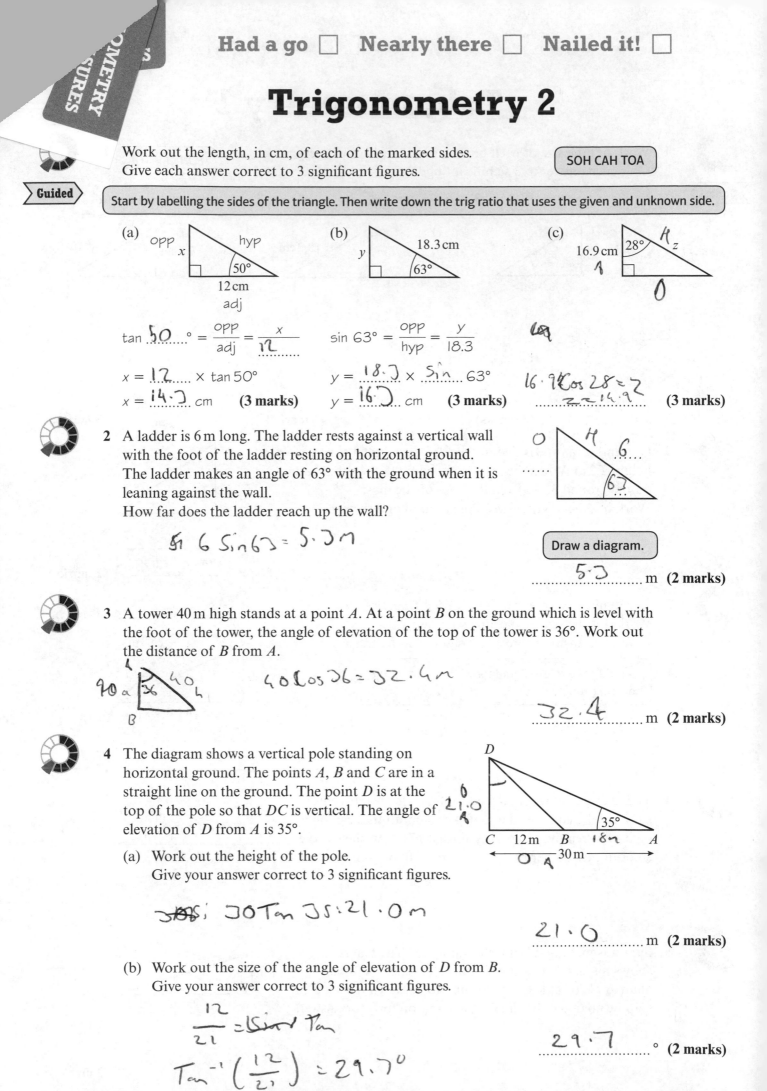

$\tan \underline{50}° = \dfrac{\text{opp}}{\text{adj}} = \dfrac{x}{12}$

$x = \underline{12} \times \tan 50°$

$x = \underline{14.3}$ cm   **(3 marks)**

(b)

$\sin 63° = \dfrac{\text{opp}}{\text{hyp}} = \dfrac{y}{18.3}$

$y = \underline{18.3} \times \underline{\sin} 63°$

$y = \underline{16.3}$ cm   **(3 marks)**

(c)

$16.9 \cos 28 = z$

$z = 14.9$ **(3 marks)**

2   A ladder is 6 m long. The ladder rests against a vertical wall with the foot of the ladder resting on horizontal ground. The ladder makes an angle of 63° with the ground when it is leaning against the wall.
How far does the ladder reach up the wall?

$6 \sin 63 = 5.3 m$

Draw a diagram.

5.3 m   **(2 marks)**

3   A tower 40 m high stands at a point A. At a point B on the ground which is level with the foot of the tower, the angle of elevation of the top of the tower is 36°. Work out the distance of B from A.

$40 \cos 36 = 32.4 m$

32.4 m   **(2 marks)**

4   The diagram shows a vertical pole standing on horizontal ground. The points A, B and C are in a straight line on the ground. The point D is at the top of the pole so that DC is vertical. The angle of elevation of D from A is 35°.

(a)   Work out the height of the pole.
Give your answer correct to 3 significant figures.

$30 \tan 35 = 21.0 m$

21.0 m   **(2 marks)**

(b)   Work out the size of the angle of elevation of D from B.
Give your answer correct to 3 significant figures.

$\dfrac{12}{21} = \text{tan}$

$\tan^{-1}\left(\dfrac{12}{21}\right) = 29.7°$

29.7 °   **(2 marks)**

Solving trigonometry problems

Guided

1 Complete the table.

You must remember these for the exam.

	0°	30°	45°	60°	90°
sin		$\frac{1}{2}$			
cos			$\frac{1}{\sqrt{2}}$		
tan				$\sqrt{3}$	

(5 marks)

2 Work out the exact length, in cm, of each of the marked sides.

SOH CAH TOA

Guided

Start by labelling the sides of the triangle. Then write down the trig ratio that uses the given and unknown side.

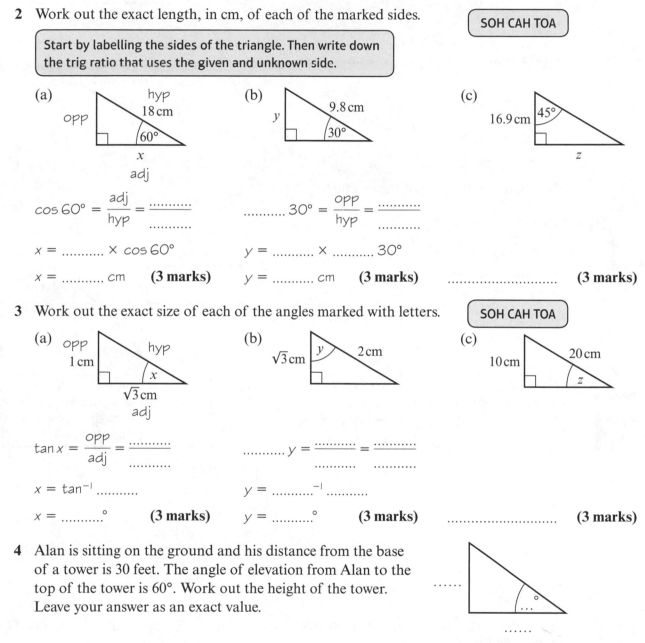

(a)

$$\cos 60° = \frac{adj}{hyp} = \frac{\dots\dots\dots}{\dots\dots\dots}$$

$$x = \dots\dots\dots \times \cos 60°$$

$$x = \dots\dots\dots \text{ cm}$$ **(3 marks)**

(b)

$$\dots\dots\dots 30° = \frac{opp}{hyp} = \frac{\dots\dots\dots}{\dots\dots\dots}$$

$$y = \dots\dots\dots \times \dots\dots\dots 30°$$

$$y = \dots\dots\dots \text{ cm}$$ **(3 marks)**

(c)

$$\dots\dots\dots\dots\dots\dots\dots\dots\dots$$ **(3 marks)**

3 Work out the exact size of each of the angles marked with letters.

SOH CAH TOA

Guided

(a)

$$\tan x = \frac{opp}{adj} = \frac{\dots\dots\dots}{\dots\dots\dots}$$

$$x = \tan^{-1} \dots\dots\dots$$

$$x = \dots\dots\dots°$$ **(3 marks)**

(b)

$$\dots\dots\dots y = \frac{\dots\dots\dots}{\dots\dots\dots} = \frac{\dots\dots\dots}{\dots\dots\dots}$$

$$y = \dots\dots\dots^{-1} \dots\dots\dots$$

$$y = \dots\dots\dots°$$ **(3 marks)**

(c)

$$\dots\dots\dots\dots\dots\dots\dots\dots\dots$$ **(3 marks)**

4 Alan is sitting on the ground and his distance from the base of a tower is 30 feet. The angle of elevation from Alan to the top of the tower is 60°. Work out the height of the tower. Leave your answer as an exact value.

Draw a diagram.

$$\dots\dots\dots\dots\dots\dots \text{ feet}$$ **(3 marks)**

5 A sailor is going to cross a plank of wood from the ground to a boat that is 3 m above the ground level. The length of the plank is 6 m. Show that the angle of elevation is 30°.

(3 marks)

Perimeter and area

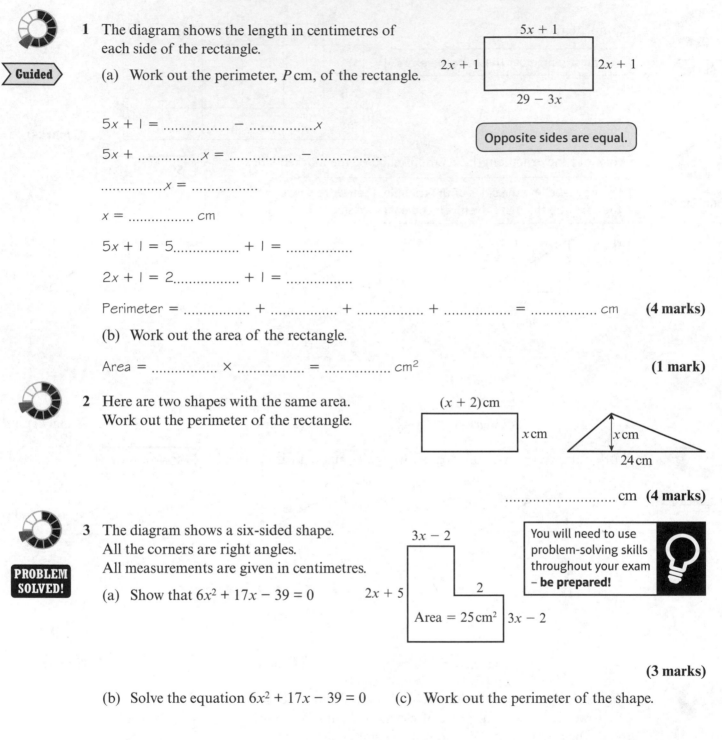

1 The diagram shows the length in centimetres of each side of the rectangle.

> **Guided**

(a) Work out the perimeter, P cm, of the rectangle.

Opposite sides are equal.

$5x + 1 =$ .... $-$ x

$5x +$ ....$x =$ .... $-$ ....

...............$x =$ ....

$x =$ .... cm

$5x + 1 = 5$............... $+ 1 =$ ....

$2x + 1 = 2$............... $+ 1 =$ ....

Perimeter = ... + ... + ... + ... = ... cm **(4 marks)**

(b) Work out the area of the rectangle.

Area = ... × .... = ... cm² **(1 mark)**

2 Here are two shapes with the same area.
Work out the perimeter of the rectangle.

$(x + 2)$ cm
x cm
x cm
24 cm

.... cm **(4 marks)**

3 The diagram shows a six-sided shape.
All the corners are right angles.
All measurements are given in centimetres.

You will need to use problem-solving skills throughout your exam – **be prepared!**

> **PROBLEM SOLVED!**

(a) Show that $6x^2 + 17x - 39 = 0$

$3x - 2$
$2x + 5$
2
Area = 25 cm² $3x - 2$

(3 marks)

(b) Solve the equation $6x^2 + 17x - 39 = 0$

.... **(3 marks)**

(c) Work out the perimeter of the shape.

.... cm **(2 marks)**

4 The length of a rectangle is twice the width of the rectangle.
The length of a diagonal of the rectangle is 25 cm.
Work out the area of the rectangle.
Give your answer as an integer.

x
$2x$

.... cm² **(3 marks)**

Units of area and volume

1 Convert

(a) 6 m² into cm²

(b) 15 cm² into mm²

(c) 4 km² into m²

$6\,m^2 = 6 \times$ .... $\times$ ....

= .... cm² **(2 marks)**

.... mm² **(2 marks)**

.... m² **(2 marks)**

(d) 500 000 cm² into m²

(e) 60 000 mm² into cm²

(f) 800 000 m² into km²

.... m² **(2 marks)**

.... cm² **(2 marks)**

.... km² **(2 marks)**

2 Convert

(a) 22 m³ into cm³

(b) 28 cm³ into mm³

(c) 3 km³ into m³

$22\,m^3 = 22 \times$ .... $\times$ .... $\times$ ....

= .... cm³ **(2 marks)**

.... mm³ **(2 marks)**

.... m³ **(2 marks)**

(d) 200 000 000 cm³ into m³

(e) 50 000 000 mm³ into cm³

(f) 420 000 000 m³ into km³

.... m³ **(2 marks)**

.... cm³ **(2 marks)**

.... km³ **(2 marks)**

3 Convert

(a) 200 000 cm³ into litres

(b) 8 m³ into litres

(c) 12 m³ into litres

.... litres **(2 marks)**

.... litres **(2 marks)**

.... litres **(2 marks)**

4 Work out how many litres of water each tank in the shape of a cuboid can hold.

(a)

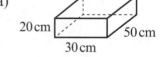

20 cm   50 cm   30 cm

(b)

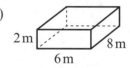

2 m   8 m   6 m

.... litres **(2 marks)**

.... litres **(2 marks)**

5 The pressure in a boiler is 18 000 N/m². The area of one end of the boiler is 4000 cm². Work out the force on the end of the boiler.

PROBLEM SOLVED!

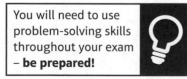

You will need to use problem-solving skills throughout your exam – **be prepared!**

.... N **(3 marks)**

Prisms

1   Find the volume of each of these prisms.

(a)

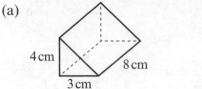

4 cm   8 cm   3 cm

> Volume of a prism = area of cross-section × length.
> You need to learn this formula for your exam.

Volume = ($\frac{1}{2}$ × .... × ....) × ....

Volume = ... cm³   **(2 marks)**

(b)

4 mm   5 mm   6 mm   12 mm

(c)

4 mm   6 mm   12 mm   8 mm

.... mm³ **(2 marks)**   .... mm³ **(2 marks)**

2   Find the surface area of the prisms in Question 1.

(a)

Surface area =

2($\frac{1}{2}$ × .... × ....) + (.......... × ....) + (.......... × ....) + (.......... × ....)

Surface area = ... cm²   **(3 marks)**

(b)   (c)

.... **(2 marks)**   .... **(3 marks)**

3   The diagram shows a prism.
The area of the cross-section of the prism is 22 cm².
The length of the prism is 16 cm.
Work out the volume of the prism.

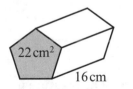

22 cm²   16 cm

.... cm³ **(2 marks)**

4   The diagram shows a triangular prism and a cuboid.
They have the same volume.
Work out the length of x.

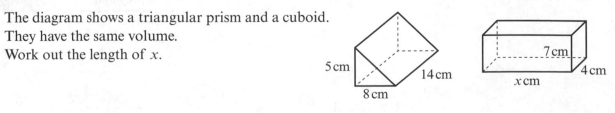

5 cm   14 cm   8 cm   7 cm x cm   4 cm

$x =$ .... cm **(4 marks)**

Circles and cylinders

1 A reel of thread has a radius of 2.5 cm. The thread is wrapped round the reel 200 times. Estimate the length of the thread.
Give your answer correct to 3 significant figures.

Guided

$C = 2 \times \pi \times r$

$C = 2 \times \pi \times$ ....

$C =$ .... cm

Length of the thread = 200 × .... = ... cm

(3 marks)

2 Bob is rolling a cricket pitch. The length of the pitch is 20.12 m.
The roller has a radius of 8 cm.
Show that the roller makes 40 complete turns.

(3 marks)

3 The diagrams show two identical squares.
Shape **A** is a quarter of a circle shaded inside the square.
Shape **B** is two identical semicircles shaded inside the square.
Show that the area of the region shaded in **A** is equal to area of the region shaded in **B**.

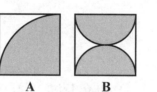

A   B

(3 marks)

4 Object **A** is a cylinder with radius 15 cm and height 18 cm.
Object **B** is a cube with side length 24 cm.
Show that the volume of the cube is greater than the volume of the cylinder.

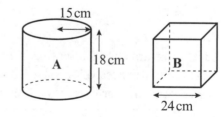

(4 marks)

5 Object **A** is a cylindrical vase with radius 9 cm and height 21 cm.
Object **B** is a vase with a square base of length 8 cm and height 42 cm.
Which vase has the greater external surface area? You must show your working.

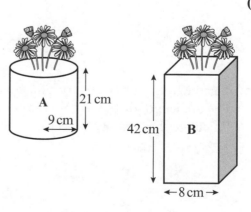

(4 marks)

Sectors of circles

1 Work out the arc lengths of these sectors of circles. Give your answers correct to 3 significant figures.

> You are only calculating the curved length in this question.

(a)

80° 4 cm

Arc length = 2 × π × r × $\dfrac{\ldots\ldots\ldots}{360}$

= 2 × π × .... × $\dfrac{\ldots\ldots\ldots}{360}$

= .... cm   **(3 marks)**

(b)

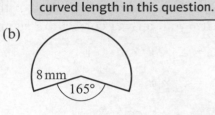

8 mm   165°

.... mm   **(3 marks)**

2 Work out the perimeters of these sectors of circles. Give your answers correct to 3 significant figures.

> Find the arc length then add the two radii.

(a)

85° 7 cm

(b)

9 mm   155°

.... cm   **(4 marks)**

.... mm   **(4 marks)**

3 Work out the areas of the sectors in Question 2. Give your answers correct to 3 significant figures.

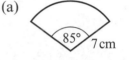

(a) $A = π × r × r × \dfrac{\ldots\ldots\ldots}{360}$

= π × .... × .... × $\dfrac{\ldots\ldots\ldots}{360}$

= .... cm²   **(3 marks)**

(b)

.... mm²   **(3 marks)**

4 The diagram shows a sector of circle with radius 10.4 cm. Calculate the area of the shaded segment *ABC*. Give your answer correct to 3 significant figures.

> Work out the area of the sector.

> Work out the area of the triangle.

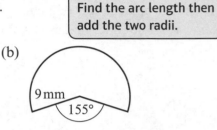

A   B   C

10.4 cm   10.4 cm

120°

O

.... cm²   **(4 marks)**

5 The diagram shows a sector of a circle, centre *O*, radius 10 cm. The arc length of the sector is 15 cm. Calculate the area of the sector.

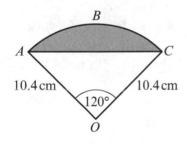

10 cm   15 cm

O   10 cm

.... cm²   **(4 marks)**

Volumes of 3D shapes

1 Work out the volumes of these solid shapes.
Give your answers correct to 3 significant figures.

Guided

(a)

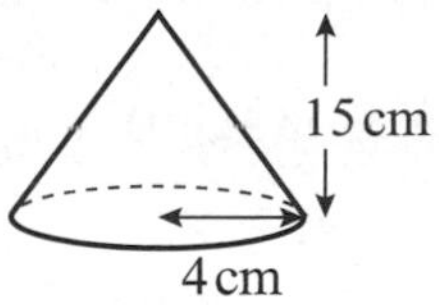
15 cm
4 cm

$V = \frac{1}{3} \times \pi \times r^2 \times h$

$= \frac{1}{3} \times \pi \times$ ....^{2} $\times$ ....

$=$ .... cm^3 **(2 marks)**

(b)

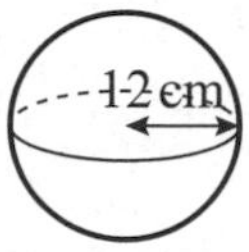
12 cm

(c)

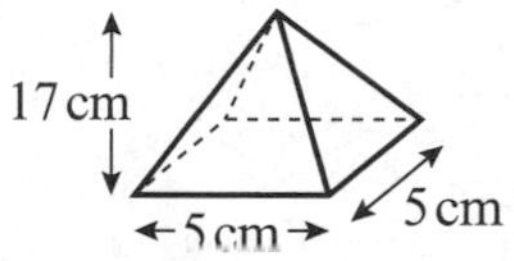
17 cm
5 cm
5 cm

.... cm^3 **(2 marks)**

.... cm^3 **(2 marks)**

2 Work out the volumes of these solid shapes. Give your answers correct to 3 significant figures.

(a)

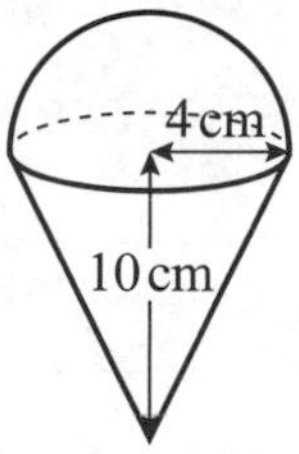
4 cm
10 cm

(b)

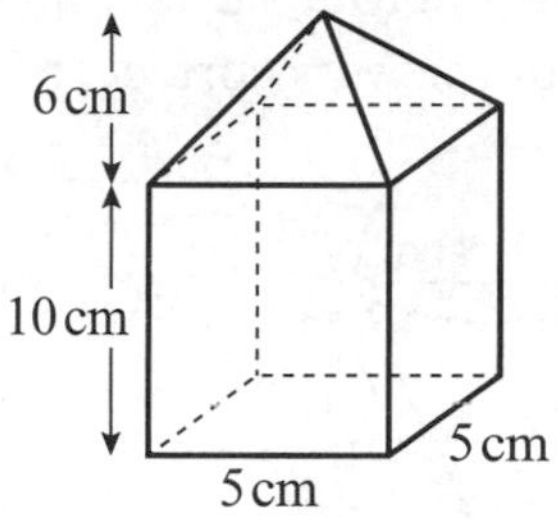
6 cm
10 cm
5 cm
5 cm

.... cm^3 **(3 marks)**

.... cm^3 **(3 marks)**

PROBLEM SOLVED!

3 A cylinder has base radius x cm and height $3x$ cm.
A cone has base radius x cm and height h cm.
The volume of the cylinder and the volume of the cone are equal.
Find h in terms of x.
Give your answer in its simplest form.

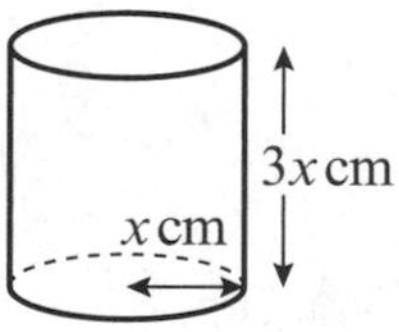
3x cm
x cm

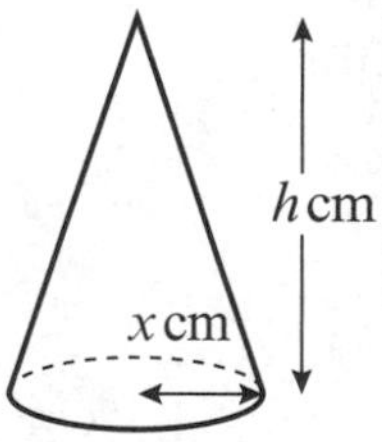
h cm
x cm

You will need to use problem-solving skills throughout your exam – **be prepared!**

$h =$ .... **(4 marks)**

4 The diagram shows a cylinder and a sphere.
The radius of the base of the cylinder is x cm and the height of the cylinder is h cm.
The radius of the sphere is x cm.
The volume of the cylinder is three times the volume of the sphere.
Show that the height, h, is 4 times the radius, x.

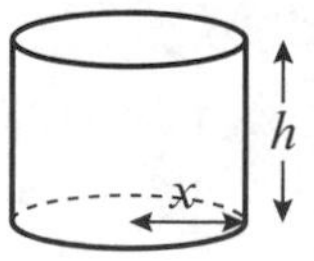
h
x

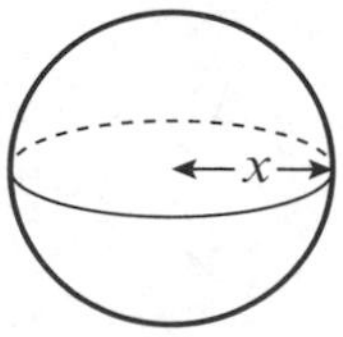
x

(4 marks)

Surface area

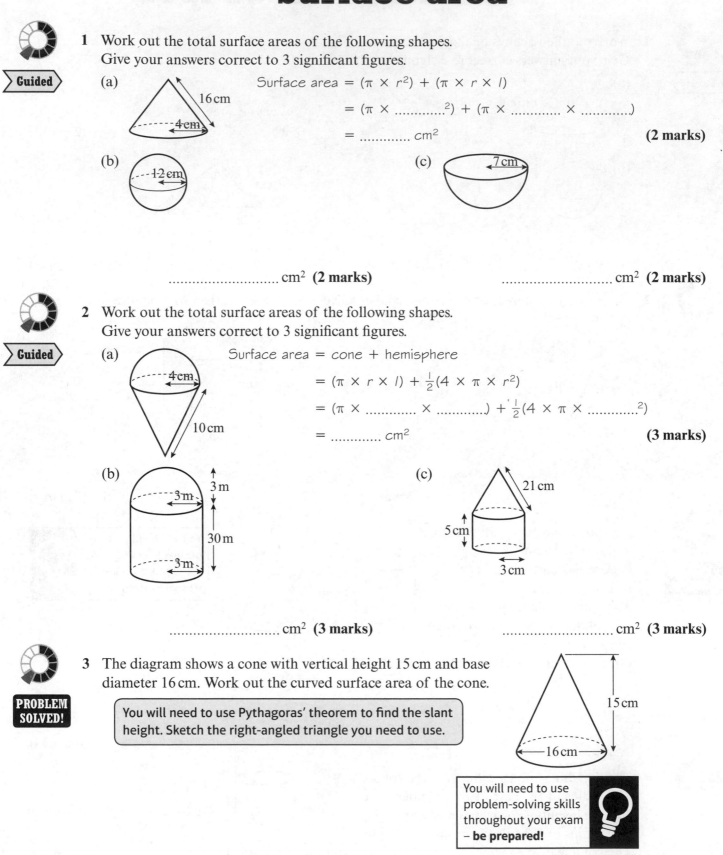

1 Work out the total surface areas of the following shapes.
Give your answers correct to 3 significant figures.

Guided

(a)

16 cm

4 cm

Surface area = $(\pi \times r^2) + (\pi \times r \times l)$

$= (\pi \times ....^2) + (\pi \times .... \times ....)$

$= ....$ cm²   **(2 marks)**

(b)

12 cm

(c)

7 cm

.... cm² **(2 marks)**   .... cm² **(2 marks)**

2 Work out the total surface areas of the following shapes.
Give your answers correct to 3 significant figures.

Guided

(a)

4 cm

10 cm

Surface area = cone + hemisphere

$= (\pi \times r \times l) + \frac{1}{2}(4 \times \pi \times r^2)$

$= (\pi \times .... \times ....) + \frac{1}{2}(4 \times \pi \times ....^2)$

$= ....$ cm²   **(3 marks)**

(b)

3 m

3 m

30 m

3 m

(c)

21 cm

5 cm

3 cm

.... cm² **(3 marks)**   .... cm² **(3 marks)**

3 The diagram shows a cone with vertical height 15 cm and base
diameter 16 cm. Work out the curved surface area of the cone.

PROBLEM SOLVED!

> You will need to use Pythagoras' theorem to find the slant
> height. Sketch the right-angled triangle you need to use.

15 cm

16 cm

> You will need to use
> problem-solving skills
> throughout your exam
> – **be prepared!**

.... cm² **(4 marks)**

Plans and elevations

1 The diagrams show the different elevations of a shape.
Using this information sketch the 3D shape.

(a)

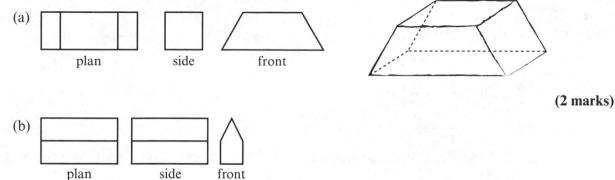

plan   side   front

(2 marks)

(b)

plan   side   front

(2 marks)

2 The diagram shows a triangular prism.
Accurately construct a plan and elevations
for this prism.

> You will need to use
> problem-solving skills
> throughout your exam
> – **be prepared!**

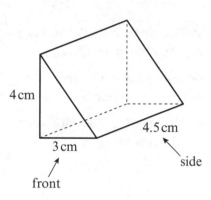

4 cm

3 cm

4.5 cm

front

side

(4 marks)

3 Here is a 3D diagram of the first two patterns in a
series of cube-shaped bricks.
Sketch a side elevation of the arrangement of
cubes of the next pattern in the series.

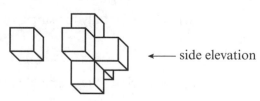

◄── side elevation

(2 marks)

Translations, reflections and rotations

1

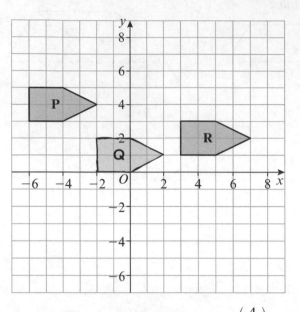

(a) Translate shape **P** by the vector $\begin{pmatrix} 4 \\ -3 \end{pmatrix}$
    Label the new shape **Q**.

(b) Reflect shape **R** in the line $y = x$
    Label the new shape **S**.

(c) Rotate shape **P** by 180° about the point $(-1, 0)$.
    Label the new shape **T**.

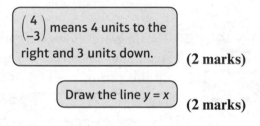

$\begin{pmatrix} 4 \\ -3 \end{pmatrix}$ means 4 units to the
right and 3 units down. **(2 marks)**

Draw the line $y = x$ **(2 marks)**

(2 marks)

2

(a) Describe fully the single transformation that will map shape **A** onto shape **B**.

.... **(2 marks)**

(b) Describe fully the single transformation that will map shape **B** onto shape **C**.

.... **(2 marks)**

(c) Describe fully the single transformation that will map shape **B** onto shape **D**.

.... **(2 marks)**

Enlargement

1 (a) Shape **B** is an enlargement of shape **A**.
   Find the scale factor of the enlargement.

Guided

.... ÷ .... = ...

15 cm

3 cm

2 cm ☐ **A**

B 10 cm

(1 mark)

(b) Enlarge the triangle below by scale factor 2.

> No centre of enlargement is given so the enlarged shape can be placed anywhere on the grid.

(1 mark)

2 Enlarge triangle **A** by scale factor 3, centre (0, 0).
   Label the new shape **B**.

Guided

> Draw lines from the centre of enlargement through each vertex of the triangle.

(2 marks)

3 Triangle **A** is shown on the grid.

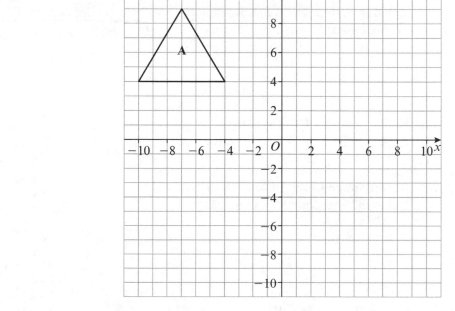

(a) Enlarge triangle **A** by scale factor $\frac{1}{2}$ with centre of enlargement (1, −3).
   Label the image **B**.
   (2 marks)

(b) Enlarge triangle **A** by scale factor $-\frac{1}{2}$ with centre of enlargement (−3, −2).
   Label the image **C**.
   (3 marks)

Combining transformations

1   (a)   Reflect shape **A** in the line $y = 1$
Label the new shape **B**.

> **Guided**

| Draw the line $y = 1$ | **(2 marks)**

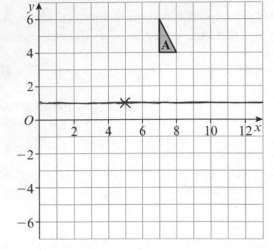

(b)   Rotate shape **B** by 180° about
the point (5, 1).
Label the new shape **C**.

| Use tracing paper to
rotate the shape. | **(2 marks)**

(c)   Describe fully the single transformation which maps shape **A** onto shape **C**.

.... **(2 marks)**

2   (a)   Translate shape **P** by the vector $\begin{pmatrix} 4 \\ 2 \end{pmatrix}$
Label the new shape **Q**.   **(2 marks)**

(b)   Rotate shape **Q** by 180° about the
point (7, 7).
Label the new shape **R**.   **(2 marks)**

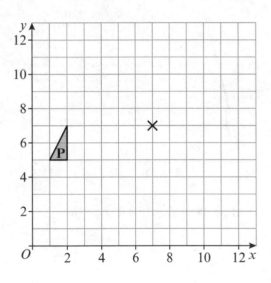

(c)   Describe fully this single
transformation which maps shape **P**
onto shape **R**.

.... **(2 marks)**

3   (a)   Reflect triangle **T** in the line $y = x$
Label the new shape **U**.   **(2 marks)**

(b)   Rotate shape **U** 180° about (0, 0).
Label the new shape **V**.   **(2 marks)**

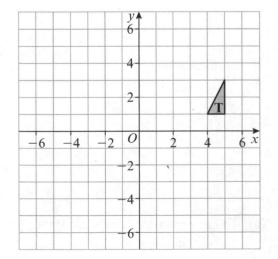

(c)   Describe fully the single
transformation which maps
shape **T** onto shape **V**.

.... **(3 marks)**

Bearings

1 The diagram shows the position of each of three telephone masts, *A*, *B* and *C*.
The bearing of mast *B* from mast A is 034°. Mast *C* is due east of mast *B*.
The distance from mast *A* to mast *B* is equal to the distance from mast *B* to mast *C*.

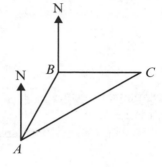

Work out the bearing of mast *C* from mast *A*.

.... **(3 marks)**

2 A submarine *S* is 3.5 km due west of a naval base *B*.
A frigate *F* is 2.4 km due north of the submarine *S*.
Find the bearing of the frigate *F* from the naval base *B*.
Give your answer correct to 3 significant figures.

> **Guided**

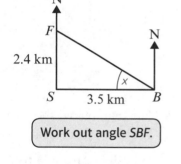

Work out angle *SBF*.

$$\tan x = \frac{\text{opp}}{\text{adj}}$$

$$\tan x = \frac{\text{.............}}{\text{.............}}$$

$$x = \tan^{-1}\frac{\text{.............}}{\text{.............}}$$

$$x = \text{.............}°$$

The bearing of the frigate *F* from the naval base *B* = ...° + ...° = ...°

(4 marks)

3 *A*, *B* and *C* are three villages.
A is 10.9 km due west of *B*.
B is 8.5 km due south of *C*.
Calculate the bearing of *A* from *C*.
Give your answer correct to 3 significant figures.

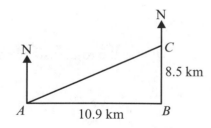

.............................° **(4 marks)**

Scale drawings and maps

Guided ▷

1   Arthur uses a scale of 1 : 300 to make a model of an aeroplane.

   (a) The wingspan of the model is 5 cm.
        Work out the wingspan of the
        aeroplane.

        300 × 5 = ...cm

                = ...m   **(2 marks)**

   (b) The length of the aeroplane is 45 m.
        Work out the length of the model.

        > The length of the model must be smaller
        > than the length of the real aeroplane.

                                    .... cm **(2 marks)**

2   Here is a scale drawing of Andrew's garden patio.
    He wants to cover the patio using slabs.
    Each slab is 50 cm by 50 cm square.

    Andrew buys 120 slabs. Does he buy enough slabs
    to completely cover his patio?
    You must show all your working.

    1 cm represents 2 m

                                                              (3 marks)

3   The map below shows a small island drawn using a scale of 1 : 500 000.
    There are three beacons at *A*, *B* and *C*.

    John drives from *A* to *B*, from *B* to *C* and from *C* back to *A*.

    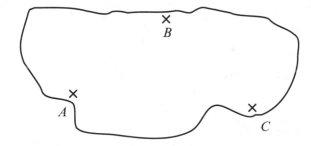

   (a) What is the total distance he drives?

                                    .... km **(3 marks)**

   (b) The bearing of *D* from *A* is 60°.
        The bearing of *D* from *C* is 290°
        Work out the distance of *D* from *B*. Give your answer in km.

        > Add point *D* on the map

                                    .... km **(3 marks)**

Constructions 1

1 Use a ruler and compasses to construct the perpendicular bisector of *AB*.

> **Guided**

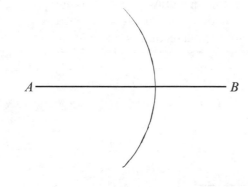

Do not rub out the arcs you make when using your compass.

1. Draw an arc, centre *A*, with radius more than half the length of *AB* above and below the line segment *AB*.
2. Draw another arc, centre *B*, with the same radius, above and below the line segment *AB*.
3. Draw a line through the two points where the arcs cross each other above and below the line segment *AB*.

(2 marks)

2 Use a ruler and compasses to construct the perpendicular to the line segment *AB* that passes through the point *T*.

> **Guided**

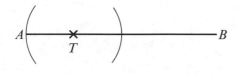

You must show all your construction lines.

First draw two arcs, centre T, with the same radius to cross *AB* on either side of *T*.

(2 marks)

3 Use a ruler and compasses to construct the perpendiculars from *P* to the line segments *AB* and *CD*. You must show all your construction lines.

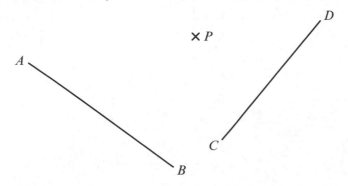

(3 marks)

Constructions 2

1 Use a ruler and compasses to construct a triangle with sides of lengths 3.5 cm, 4 cm and 5 cm.

> Do not rub out the arcs you draw when using your compasses.

> 1. Draw a horizontal line of 5 cm and label it *AB*.
> 2. Set the compasses at 3.5 cm, then draw an arc with centre *A*.
> 3. Set the compasses at 4 cm, then draw an arc with centre *B*.
> 4. Draw lines from the point of intersection to *A* and *B*.

A ——————5 cm————— B

(2 marks)

2 Use a ruler and compasses to construct the bisector of angle *ABC*.

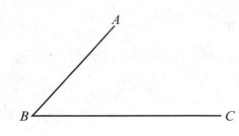

> 1. Draw an arc, centre *B*, to cross *AB* at *P* and *BC* at *Q*.
> 2. Draw two arcs, centres *P* and *Q*, with the same radius. The two arcs cross.
> 3. Draw a line through where the arcs cross to *B*.

(2 marks)

3 Use a ruler and compasses to construct a 60° angle at *A*. You must show all your construction lines.

A ————————— B

(2 marks)

4 Use a ruler and compasses to construct a 45° angle at *A*. You must show all your construction lines.

> Start by constructing a perpendicular to *AB* through *A*.

A ————————— B

(2 marks)

Loci

 1 Draw the locus of all points that are exactly 2 cm from the line AB.

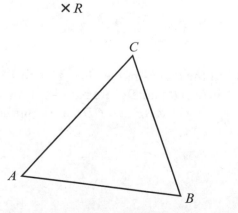
> 1. Draw a circle of radius 2 cm with centre A.
> 2. Draw a circle of radius 2 cm with centre B.
> 3. Draw two parallel lines 2 cm above and below the line AB.

A ——————— B

(2 marks)

2 The diagram shows the boundary of a rectangular garden, $ABCD$. A dog is tied to the corner C with a rope of length 6 m. Shade the region where the dog can reach.

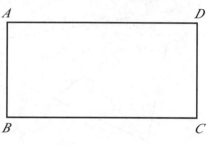

> 1. Use the scale to set the compasses at the required distance.
> 2. Draw an arc with centre C.
> 3. Shade the required region.

1 cm represents 2 m

(2 marks)

3 P, Q and R represent three radio masts on a map. Signals from mast P can be received 100 km away, and from masts Q and R 75 km away. Shade the region in which signals can be received from all three masts.

1 cm represents 25 km

$\times Q$

$P \times$

$\times R$ **(3 marks)**

4 ABC is a triangle. Shade the region inside the triangle which is both less than 3 cm from B and closer to the line AC than the line AB.

C

A

B **(3 marks)**

Congruent triangles

1 Show that triangle *ABD* is congruent to triangle *CDB*.

Guided

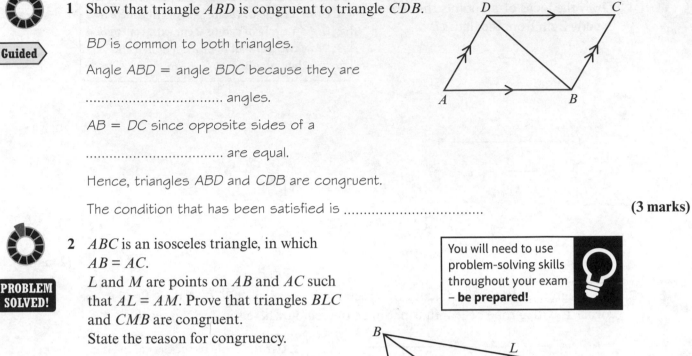

BD is common to both triangles.

Angle ABD = angle BDC because they are

.... angles.

AB = DC since opposite sides of a

.... are equal.

Hence, triangles ABD and CDB are congruent.

The condition that has been satisfied is ....

(3 marks)

2 *ABC* is an isosceles triangle, in which
AB = *AC*.
L and *M* are points on *AB* and *AC* such
that *AL* = *AM*. Prove that triangles *BLC*
and *CMB* are congruent.
State the reason for congruency.

> You will need to use
> problem-solving skills
> throughout your exam
> **– be prepared!**

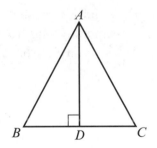

(3 marks)

3 *ABC* is an equilateral triangle. *D* lies on *BC*.
AD is perpendicular to *BC*.

(a) Prove that triangle *ADC* is congruent to triangle *ADB*.

(3 marks)

(b) Hence, prove that $DC = \frac{1}{2}AC$.

(2 marks)

4 In the diagram, the lines *AC* and *BD* intersect
at *E*. *AB* and *DC* are parallel. Prove that
triangles *ABE* and *CDE* are congruent.

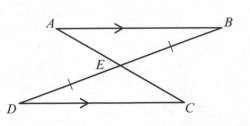

(4 marks)

Similar shapes 1

 Guided

1 The two triangles ABC and PQR are mathematically similar.

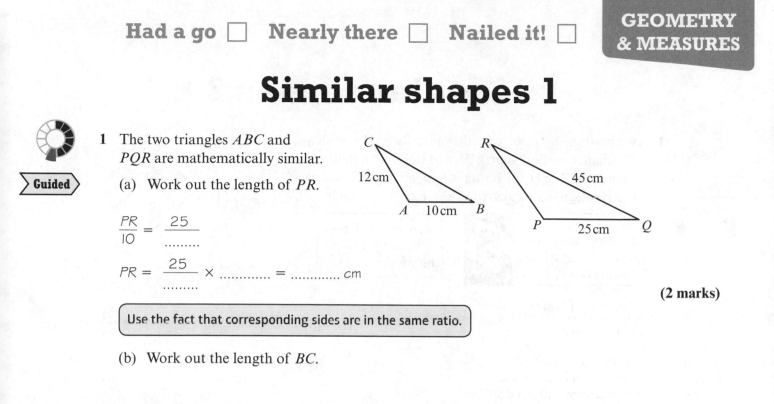

(a) Work out the length of PR.

$$\frac{PR}{10} = \frac{25}{\text{.........}}$$

$$PR = \frac{25}{\text{.........}} \times \text{............} = \text{............ } cm$$

(2 marks)

> Use the fact that corresponding sides are in the same ratio.

(b) Work out the length of BC.

.... cm **(2 marks)**

2 Triangle ABC is mathematically similar to triangle ADE. Work out the length of DE.

> Separate into two triangles.

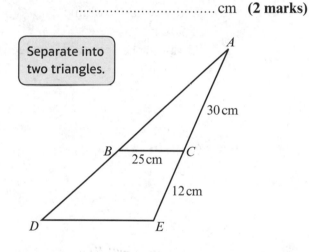

.... cm **(3 marks)**

3 Triangles ABC and ACD are mathematically similar.
$AB = 8$ cm and $AC = 10$ cm.
Calculate the length of AD.

.... cm **(3 marks)**

4 AB is parallel to DC. The lines AC and BD intersect at E. $AB = 6$ cm, $AE = 8$ cm, $DE = 14$ cm, $EC = 20$ cm. Work out the length of BE.

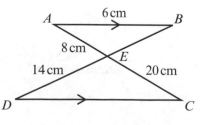

.... cm **(3 marks)**

Similar shapes 2

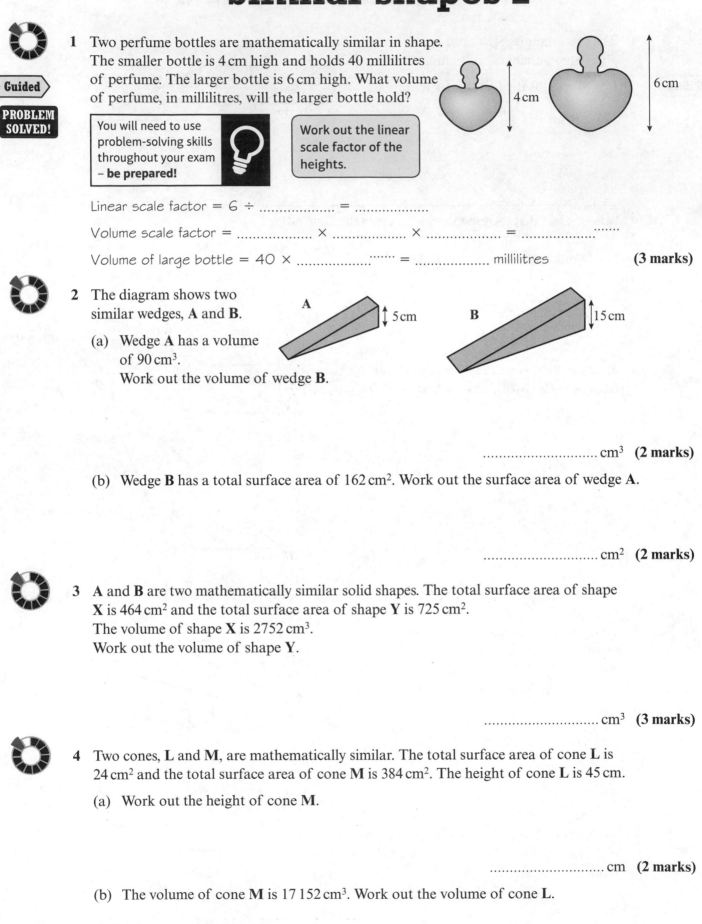

Guided

PROBLEM SOLVED!

1 Two perfume bottles are mathematically similar in shape. The smaller bottle is 4 cm high and holds 40 millilitres of perfume. The larger bottle is 6 cm high. What volume of perfume, in millilitres, will the larger bottle hold?

You will need to use problem-solving skills throughout your exam – **be prepared!**

Work out the linear scale factor of the heights.

4 cm

6 cm

Linear scale factor = 6 ÷ .... = ...

Volume scale factor = ... × .... × .... = ...

Volume of large bottle = 40 × .... = ... millilitres   **(3 marks)**

2 The diagram shows two similar wedges, **A** and **B**.

A   ↕ 5 cm   B   ↕ 15 cm

(a) Wedge **A** has a volume of 90 cm³.
Work out the volume of wedge **B**.

.... cm³ **(2 marks)**

(b) Wedge **B** has a total surface area of 162 cm². Work out the surface area of wedge **A**.

.... cm² **(2 marks)**

3 **A** and **B** are two mathematically similar solid shapes. The total surface area of shape **X** is 464 cm² and the total surface area of shape **Y** is 725 cm².
The volume of shape **X** is 2752 cm³.
Work out the volume of shape **Y**.

.... cm³ **(3 marks)**

4 Two cones, **L** and **M**, are mathematically similar. The total surface area of cone **L** is 24 cm² and the total surface area of cone **M** is 384 cm². The height of cone **L** is 45 cm.

(a) Work out the height of cone **M**.

.... cm **(2 marks)**

(b) The volume of cone **M** is 17 152 cm³. Work out the volume of cone **L**.

.... cm³ **(2 marks)**

The sine rule

1 Find the lengths of the sides marked with a letter.

Guided

(a)

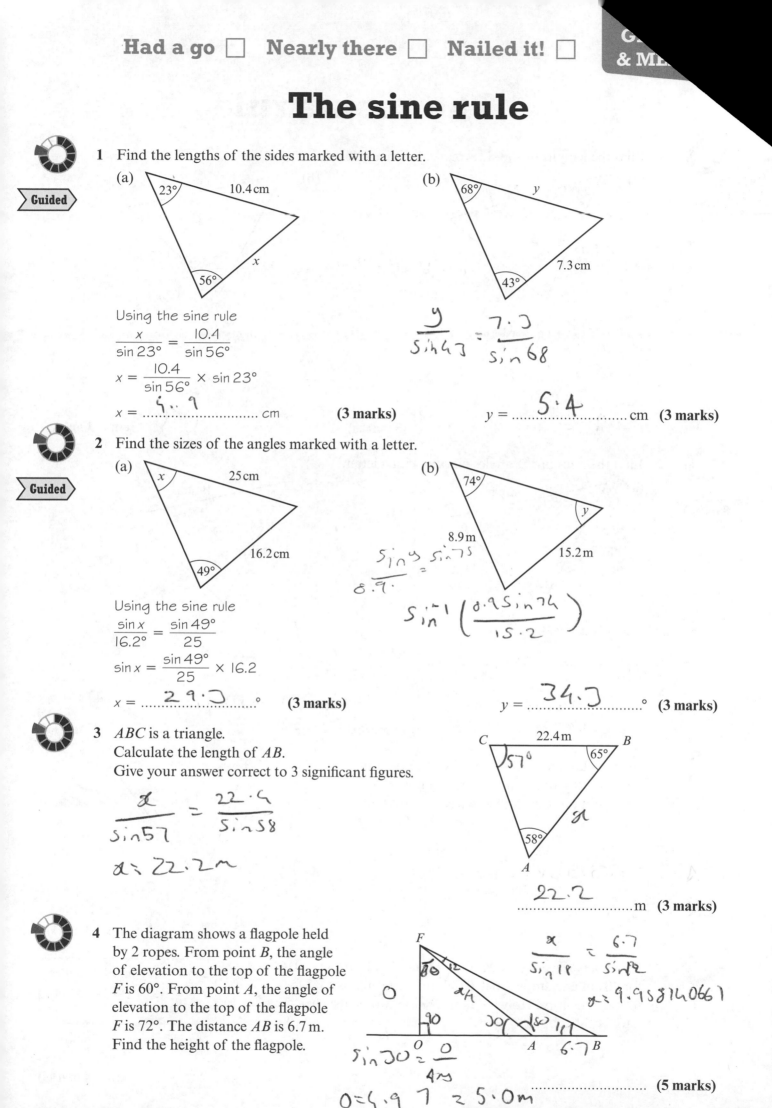

Using the sine rule

$$\frac{x}{\sin 23°} = \frac{10.4}{\sin 56°}$$

$$x = \frac{10.4}{\sin 56°} \times \sin 23°$$

$x =$ ....5..9....... cm   **(3 marks)**

(b)

$$\frac{y}{\sin 43} = \frac{7.3}{\sin 68}$$

$y =$ ....5·4....... cm   **(3 marks)**

2 Find the sizes of the angles marked with a letter.

(a)

Using the sine rule

$$\frac{\sin x}{16.2°} = \frac{\sin 49°}{25}$$

$$\sin x = \frac{\sin 49°}{25} \times 16.2$$

$x =$ ....29·3....... °   **(3 marks)**

(b)

$$\frac{\sin y}{8.9} = \frac{\sin 75}{}$$

$$\sin^{-1}\left(\frac{0.9 \sin 74}{15.2}\right)$$

$y =$ ....34·3....... °   **(3 marks)**

3 ABC is a triangle.
Calculate the length of AB.
Give your answer correct to 3 significant figures.

$$\frac{x}{\sin 57} = \frac{22.4}{\sin 58}$$

$$x = 22.2m$$

.......22·2....... m   **(3 marks)**

4 The diagram shows a flagpole held
by 2 ropes. From point B, the angle
of elevation to the top of the flagpole
F is 60°. From point A, the angle of
elevation to the top of the flagpole
F is 72°. The distance AB is 6.7 m.
Find the height of the flagpole.

$$\frac{x}{\sin 18} = \frac{6.7}{\sin 12}$$

$$x = 9.958146061$$

$$\sin 30 = \frac{0}{4x}$$

$$0 = 9.9 \quad 7 = 5.0m$$

.... m   **(5 marks)**

The cosine rule

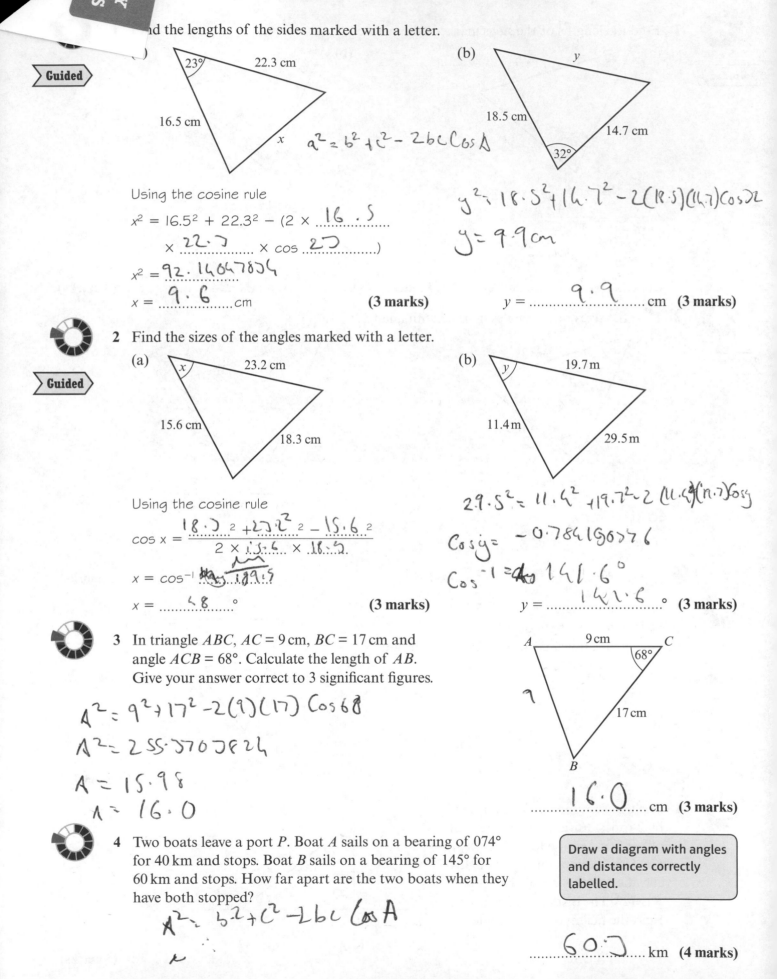

nd the lengths of the sides marked with a letter.

> **Guided**

(a)

23° 22.3 cm

16.5 cm

x

$a^2 = b^2 + c^2 - 2bc \cos A$

Using the cosine rule

$x^2 = 16.5^2 + 22.3^2 - (2 \times \underline{16.5}$

$\times \underline{22.3} \times \cos \underline{23})$

$x^2 = \underline{92.14647806}$

$x = \underline{9.6}$ cm   **(3 marks)**

(b)

y

18.5 cm

14.7 cm

32°

$y^2 = 18.5^2 + 14.7^2 - 2(18.5)(14.7) \cos 32$

$y = 9.9$ cm

$y = \underline{9.9}$ cm   **(3 marks)**

2 Find the sizes of the angles marked with a letter.

(a)

> **Guided**

x 23.2 cm

15.6 cm

18.3 cm

Using the cosine rule

$\cos x = \dfrac{18.3^2 + 23.2^2 - 15.6^2}{2 \times 15.6 \times 18.3}$

$x = \cos^{-1} \underline{189.5}$

$x = \underline{68}°$ **(3 marks)**

(b)

y 19.7 m

11.4 m

29.5 m

$29.5^2 = 11.4^2 + 19.7^2 - 2(11.4)(19.7) \cos y$

$\cos y = -0.784196076$

$\cos^{-1} = 141.6°$

$y = \underline{141.6}°$ **(3 marks)**

3 In triangle ABC, $AC = 9$ cm, $BC = 17$ cm and angle $ACB = 68°$. Calculate the length of AB. Give your answer correct to 3 significant figures.

$A^2 = 9^2 + 17^2 - 2(9)(17) \cos 68$

$A^2 = 255.370582b$

$A = 15.98$

$A = 16.0$

A 9 cm C

68°

a

17 cm

B

$\underline{16.0}$ cm   **(3 marks)**

4 Two boats leave a port P. Boat A sails on a bearing of 074° for 40 km and stops. Boat B sails on a bearing of 145° for 60 km and stops. How far apart are the two boats when they have both stopped?

$A^2 = b^2 + c^2 - 2bc \cos A$

Draw a diagram with angles and distances correctly labelled.

$\underline{60.3}$ km   **(4 marks)**

Triangles and segments

1 Work out the areas of the triangles.

Guided

(a)

23° ··· 22.3 cm

16.5 cm

Using the formula for the area

Area $= \frac{1}{2}ab \sin C$

$= \frac{1}{2} \times \text{.........} \times \text{.........} \times \sin \text{.........}$

$= \text{.............} cm^2$ **(2 marks)**

(b)

18.5 cm

32°   14.7 cm

.... cm² **(2 marks)**

2 Work out the areas of the shaded regions.

Guided

(a)

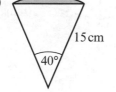

15 cm

40°

Using the formula for the area of a triangle

Area $= \frac{1}{2}ab \sin C$

$= \frac{1}{2} \times \text{.........} \times \text{.........} \times \sin \text{.........}$

$= \text{.............} cm^2$

Using the formula for the area of a sector

Area $= \pi r^2 \times \dfrac{\theta}{360}$

$= \pi \times \text{.........}^2 \times \dfrac{\text{.........}}{360} = \text{.........} cm^2$

Shaded area = Area of sector − Area of triangle

$= \text{.........} - \text{.........} = \text{.........} \; cm^2$ **(5 marks)**

(b)

105°   23 m

.... m² **(5 marks)**

3 The diagram shows an equilateral triangle *ABC* with sides of length 10 m. *D* is the midpoint of *AB*. *E* is the midpoint of *AC*. *ADE* is a sector of a circle with centre *A*.

(a) Work out the area of the triangle *ABC*.
  Give your answer correct to 3 significant figures.

.... m² **(2 marks)**

(b) Hence, or otherwise, work out the area of the shaded region.
  Give your answer correct to 3 significant figures.

.... m² **(3 marks)**

Pythagoras in 3D

Guided

1 A cuboid has length 3 cm, width 4 cm and height 12 cm.
   Work out the length of *PQ*.

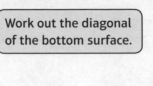

Work out the diagonal
of the bottom surface.

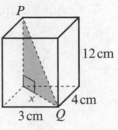

$x^2 = $ ....² + ...²

$x = \sqrt{\rule{2cm}{0pt}}$

$x = $ .... cm

$PQ^2 = $ ....² + ...²

$PQ = $ .... cm

(3 marks)

2 The diagram represents a cuboid *ABCDEFGH*.
   AB = 8 cm, *BC* = 10 cm, *AE* = 3.5 cm.

   Calculate the length of *AG*. Give your
   answer correct to 3 significant figures.

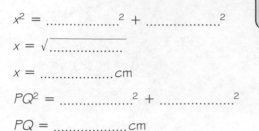

.... cm **(3 marks)**

3 The diagram shows a pyramid. The apex of the
   pyramid is *V*. Each of the sloping sides of the
   pyramid is 8 cm. The base of the pyramid is a
   regular hexagon with sides of length 3 cm.
   O is the centre of the base.

   Calculate the height of *V* above the base of
   the pyramid. Give your answer correct to
   3 significant figures.

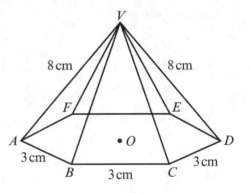

.... cm **(4 marks)**

4 A box has length 30 cm, width 8 cm and height 11 cm.
   Show that a stick of length of 35 cm will not fit completely in the box.

(4 marks)

Trigonometry in 3D

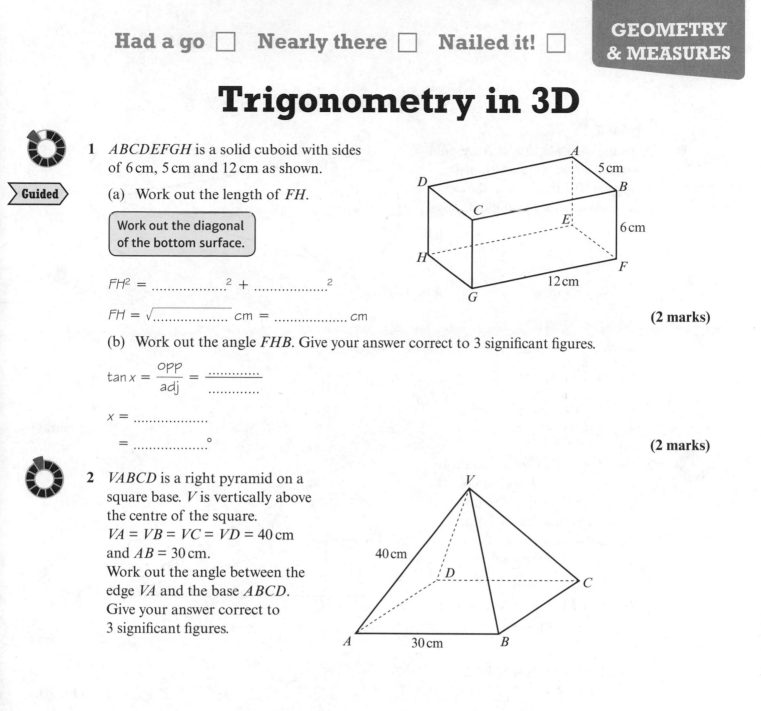

Guided

1   *ABCDEFGH* is a solid cuboid with sides of 6 cm, 5 cm and 12 cm as shown.

(a) Work out the length of *FH*.

> Work out the diagonal of the bottom surface.

$FH^2 = \underline{\hspace{1.5cm}}^2 + \underline{\hspace{1.5cm}}^2$

$FH = \sqrt{\underline{\hspace{1.5cm}}}$ cm $= \underline{\hspace{1.5cm}}$ cm   **(2 marks)**

(b) Work out the angle *FHB*. Give your answer correct to 3 significant figures.

$\tan x = \dfrac{opp}{adj} = \dfrac{\underline{\hspace{1cm}}}{\underline{\hspace{1cm}}}$

$x = \underline{\hspace{2cm}}$

$= \underline{\hspace{2cm}}°$ **(2 marks)**

2   *VABCD* is a right pyramid on a square base. *V* is vertically above the centre of the square.
VA = VB = VC = VD = 40 cm and *AB* = 30 cm.
Work out the angle between the edge *VA* and the base *ABCD*.
Give your answer correct to 3 significant figures.

$\underline{\hspace{4cm}}°$ **(4 marks)**

3   The diagram shows a door wedge with a rectangular horizontal base *PQRS* and a rectangular sloping face *PQTU*.
PQ = 4.2 cm and angle *TQR* = 12°.
The height *TR* is 3 cm.
Calculate the length of the diagonal *PT*.
Give your answer correct to 3 significant figures.

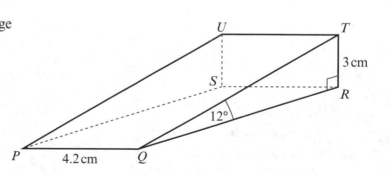

$\underline{\hspace{4cm}}$ cm **(4 marks)**

Circle facts

1 In the diagram, *A* and *B* are points on the circle. *AC* and *BC* are tangents to the circle and meet at *C*. Work out the angle *ACB*.

> **Guided**

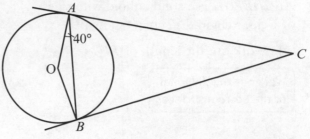

Triangle *AOB* is an .... triangle.

Angle *AOB* = 180° − (.........................° + ...°) = ...°

Angles in the quadrilateral *OACB* add up to ....°

Angle *OAC* = Angle *OBC* = ...°

Angle *ACB* = ...° − (.........................° + ...° + ...°)

Angle *ACB* = ...°

(3 marks)

2 The diagram shows a circle with centre *O*. *A*, *L*, *B* and *M* are points on the circumference of the circle. *MN* and *LN* are tangents to the circle. *AB* is parallel to *MN* and angle *LNM* is 48°.

(a) (i) Write down the size of angle *OLN*.

.........................° **(1 mark)**

(ii) Give a reason for your answer.

.... **(1 mark)**

(b) Work out the size of angle *LOB*.

.........................° **(2 marks)**

3 *A* and *B* are two points on a circle with centre *O*. *BN* is a tangent. Angle *ABN* is *x*. Prove that the size of angle *AOB* is 2*x*.

(4 marks)

Circle theorems

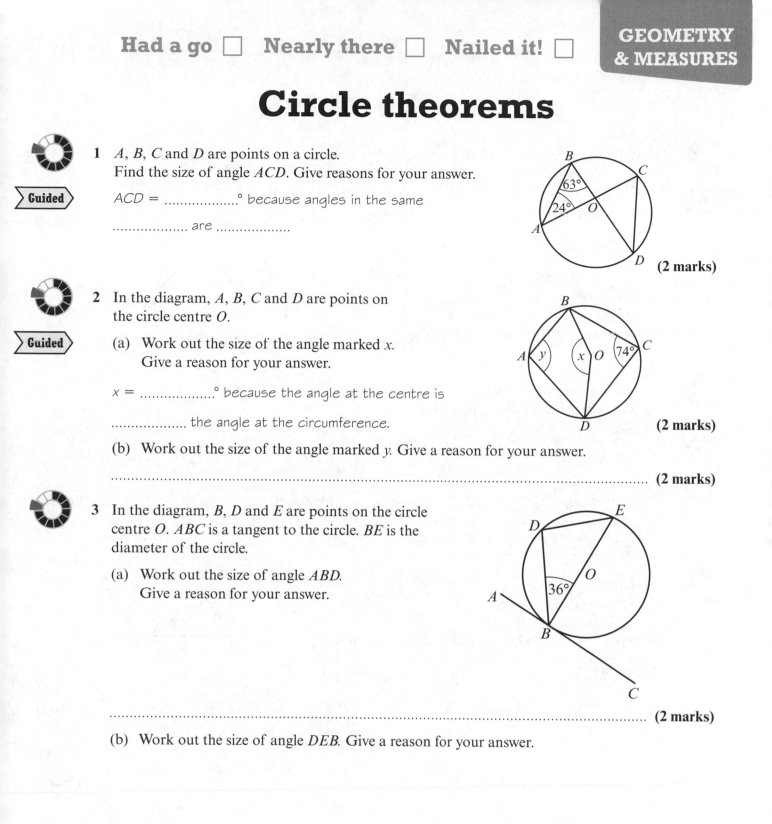

1 *A*, *B*, *C* and *D* are points on a circle.
Find the size of angle *ACD*. Give reasons for your answer.

> **Guided**

ACD = ...° because angles in the same

.... are ....

(2 marks)

2 In the diagram, *A*, *B*, *C* and *D* are points on the circle centre *O*.

> **Guided**

(a) Work out the size of the angle marked *x*.
Give a reason for your answer.

x = ...° because the angle at the centre is

.... the angle at the circumference.

(2 marks)

(b) Work out the size of the angle marked *y*. Give a reason for your answer.

.... (2 marks)

3 In the diagram, *B*, *D* and *E* are points on the circle centre *O*. *ABC* is a tangent to the circle. *BE* is the diameter of the circle.

(a) Work out the size of angle *ABD*.
Give a reason for your answer.

.... (2 marks)

(b) Work out the size of angle *DEB*. Give a reason for your answer.

.... (2 marks)

4 *A*, *B* and *C* are points on the circumference of the circle centre *O*.
The line *XCY* is the tangent at *C* to the circle. *AB* = *CB*.
Work out the size of angle *OCB*.
Give reasons for your answer.

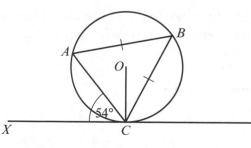

.... (5 marks)

Vectors

1 Write each vector as a column vector.

(a) $\overrightarrow{AB} = \begin{pmatrix} 2 \\ \end{pmatrix}$
(1 mark)

(b) $\overrightarrow{BA} = \begin{pmatrix} \\ -5 \end{pmatrix}$
(1 mark)

(c) $\overrightarrow{CD} = \begin{pmatrix} \\ \end{pmatrix}$
(1 mark)

(d) $\overrightarrow{DC} = \begin{pmatrix} \\ \end{pmatrix}$
(1 mark)

(e) $\overrightarrow{EF} = \begin{pmatrix} \\ \end{pmatrix}$
(1 mark)

(f) $\overrightarrow{FE} = \begin{pmatrix} \\ \end{pmatrix}$
(1 mark)

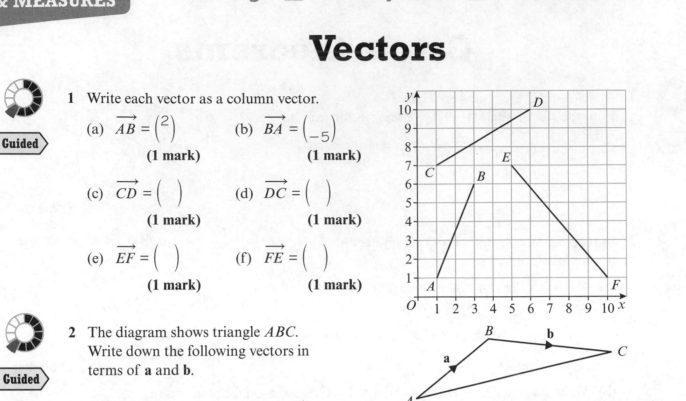

2 The diagram shows triangle *ABC*. Write down the following vectors in terms of **a** and **b**.

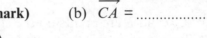

(a) $\overrightarrow{AC} = \mathbf{a} +$ .... **(1 mark)**

(b) $\overrightarrow{CA} =$ .... **(1 mark)**

> Underline any vectors you write down.

3 *ABCD* is a parallelogram.
AB is parallel to *DC*.
AD is parallel to *BC*.
Write down the following vectors in terms of **p** and **q**.

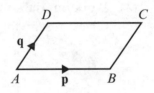

(a) $\overrightarrow{AC}$

.... **(1 mark)**

(b) $\overrightarrow{CA}$

.... **(1 mark)**

(c) $\overrightarrow{DB}$

.... **(1 mark)**

(d) $\overrightarrow{BD}$

.... **(1 mark)**

4 $\overrightarrow{OB} = 3\mathbf{a} + 2\mathbf{b}$

$\overrightarrow{OA} = 2\mathbf{a} + \mathbf{b}$

Write down the following vectors in terms of **a** and **b**.
Give your answers in their simplest form.

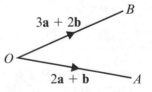

(a) $\overrightarrow{AB}$

.... **(2 marks)**

(b) $\overrightarrow{BA}$

.... **(2 marks)**

Vector proof

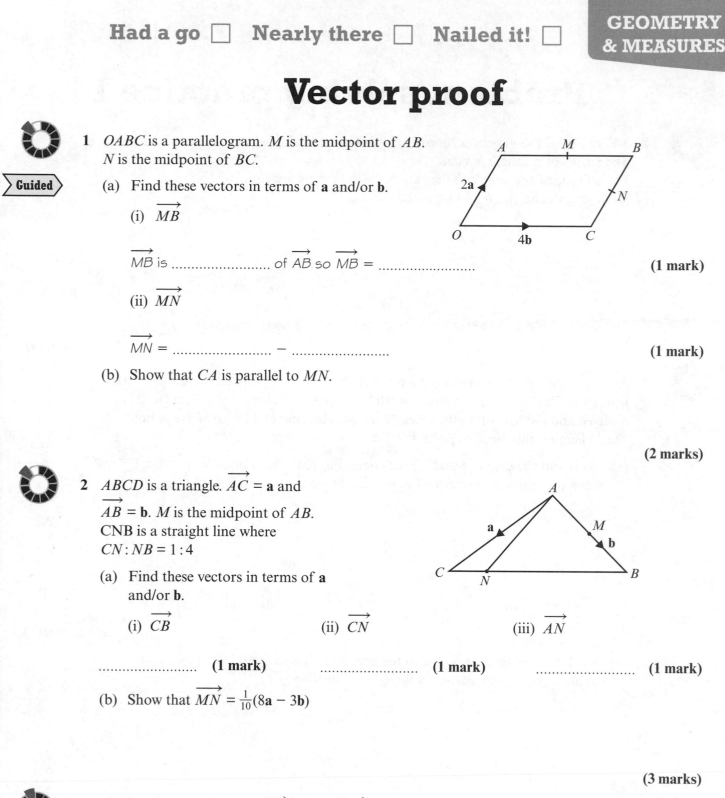

1 *OABC* is a parallelogram. *M* is the midpoint of *AB*.
N is the midpoint of *BC*.

Guided

(a) Find these vectors in terms of **a** and/or **b**.

(i) $\overrightarrow{MB}$

$\overrightarrow{MB}$ is .... of $\overrightarrow{AB}$ so $\overrightarrow{MB}$ = ... **(1 mark)**

(ii) $\overrightarrow{MN}$

$\overrightarrow{MN}$ = ... − .... **(1 mark)**

(b) Show that *CA* is parallel to *MN*.

(2 marks)

2 *ABCD* is a triangle. $\overrightarrow{AC}$ = **a** and
$\overrightarrow{AB}$ = **b**. *M* is the midpoint of *AB*.
CNB is a straight line where
CN : *NB* = 1 : 4

(a) Find these vectors in terms of **a**
and/or **b**.

(i) $\overrightarrow{CB}$ (ii) $\overrightarrow{CN}$ (iii) $\overrightarrow{AN}$

.... **(1 mark)**   .... **(1 mark)**   .... **(1 mark)**

(b) Show that $\overrightarrow{MN} = \frac{1}{10}(8\mathbf{a} - 3\mathbf{b})$

(3 marks)

3 *OABC* is a parallelogram. $\overrightarrow{OA}$ = **a** and $\overrightarrow{OD}$ = **c**.
BCE is a straight line where *BC* : *CE* = 1 : 2
D is the midpoint of *OC*.

Work out the ratio of the lengths *AD* : *OE*

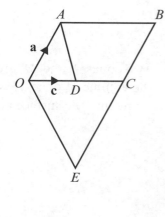

.... **(5 marks)**

Problem-solving practice 1

1 The radius of the base of a cone is x cm and its vertical height is h cm.
The radius of a sphere is x cm.
The volume of the cone and the volume of the sphere are equal.
Show that the height, h, is 4 times the radius, x.

(4 marks)

2 A rescue helicopter is stationed at a point X. An emergency call is received and the helicopter flies 16 km on a bearing of 040° to point Y. A second emergency call is received and the helicopter then flies 30 km on a bearing of 115° to arrive at point Z. The helicopter flies back to point X.

(a) Work out the distance that the helicopter has to fly from point Z to point X.
Give your answer correct to 3 significant figures.

.... km **(3 marks)**

(b) Work out the bearing on which the helicopter has to fly to from point Z to point X. Give your answer as three-figure bearing.

...........................° **(3 marks)**

3 Two cones, L and M, are mathematically similar. The ratio of the volume of cone L to the volume of cone M is $27 : 125$. The surface area of cone M is 450 cm². Show that the surface area of cone L is 162 cm².

(3 marks)

Problem-solving practice 2

4 The diagram shows a pyramid with a horizontal rectangular base $PQRS$. $PQ = 18$ cm, $QR = 12$ cm and $MT = 16$ cm. M is the midpoint of the line PR. The vertex T is vertically above M.

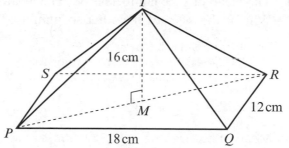

Work out the size of angle between TP and the base $PQRS$.

Give your answer correct to 3 significant figures.

..............................° **(4 marks)**

5 $ABCD$ is a trapezium.
AB is parallel to DC.
E is the point on the diagonal such that $DE = \frac{1}{4}DB$.

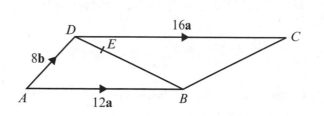

Prove that BC is parallel to AE.

(5 marks)

6 AOD is the diameter of a circle with centre O and radius 12 cm. ABC is an arc of the circle. Calculate the area of the shaded segment. Give your answer correct to 3 significant figures.

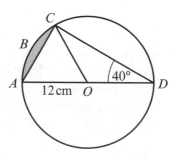

..............................cm^2 **(5 marks)**

Mean, median and mode

1 The mean of nine numbers is 62. The mean of three numbers is 52.
What is the mean of the other six numbers?

Guided

Total of nine numbers = 9 × 62 = ...

Total of three numbers = ... × .... = ...

Difference = ... – .... = ...

Mean of six numbers = ... = ... **(3 marks)**

PROBLEM SOLVED!

2 Emma has five cards. She wants to write
down a number on each card such that
- the mode of the five numbers is 7
- the median of the five numbers is 8
- the mean of the five numbers is 9
- the range of the five numbers is 5.

> You will need to use
> problem-solving skills
> throughout your exam
> – **be prepared!**

[?] [?] [?] [?] [?]

Work out the five numbers on the cards.

.... **(3 marks)**

3 There are five cards with numbers
written on them.
X is a prime number and Y is a
square number.
The five cards have a mean of 11.
Find X and Y.

[12] [8] [17] [X] [Y]

$X = $ ....

$Y = $ .... **(3 marks)**

4 Jacqui writes down the length, in cm, of
plants on a piece of paper. She accidentally
rips off the last result. Jacqui states that the
mean, the median and the mode are all equal.
Work out the value of the missing number.

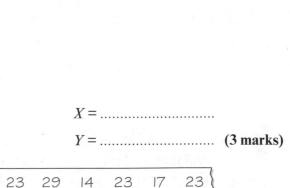

23 29 14 23 17 23

.... **(3 marks)**

Frequency table averages

Guided

1 The table shows the numbers of goals scored by a football team in each of 30 matches.

Number of goals	Frequency	
0	7	0 × 7 = ...
1	9	1 × 9 = ...
2	6	2 × 6 = ...
3	5	3 × 5 = ...
4	3	4 × 3 = ...

Draw an extra column and an extra row.

Add up the final column to work out the total number of goals.

Work out

(a) the mode

Mode is ....   **(1 mark)**

(b) the median

Median = $\dfrac{30 + 1}{2}$th value = ... value   Median = ... **(2 marks)**

(c) the mean

Mean = $\dfrac{\text{total number of goals}}{\text{total frequency}}$ = $\dfrac{...........................}{...........................}$ Mean = ... **(3 marks)**

(d) the range

Range = highest value − lowest value = ... − .... = ... **(2 marks)**

Guided

2 The table shows information about the number of hours spent on the internet last week by a group of students.

Number of hours	Frequency, f	Midpoint, x	$f \times x$
$0 \leqslant h < 2$	6	1	1 × 6 = ...
$2 \leqslant h < 4$	7	3	3 × 7 = ...
$4 \leqslant h < 6$	3	5	5 × 3 = ...
$6 \leqslant h < 8$	9	7	7 × 9 = ...
$8 \leqslant h < 10$	10	9	9 × 10 = ...

Multiply the frequency by the midpoint of each group.

(a) Work out an estimate for the mean number of hours.

Mean = $\dfrac{\text{total number of hours}}{\text{total frequency}}$ = $\dfrac{...........................}{...........................}$

Mean = ...

Add up the final column to work out the total number of hours.   **(4 marks)**

(b) Explain why your answer to part (a) is an estimate.

.... **(1 mark)**

Interquartile range

1 Roseanna recorded the masses, in kg, of 15 people. Here are her results.

45 47 50 51 57 59 62 63 65 67 70 73 76 77 78

Guided

(a) Work out the range.

Range = highest value − lowest value = ... − .... = ... **(1 mark)**

(b) Work out the interquartile range.

$\frac{1}{4}(n + 1) = \frac{1}{4}($................. $+ 1) = $.............................

So $Q_1 = $.................th value = ...

$\frac{3}{4}(n + 1) = \frac{3}{4}($................. $+ 1) = $.............................

So $Q_3 = $.................th value = ...

Interquartile range = $Q_3 − Q_1 = $..................... − .... = ... **(2 marks)**

2 The stem-and-leaf diagram shows information about the ages, in years, of some people at a shop.

(a) Work out the range.

.... years **(1 mark)**

(b) Work out the interquartile range

0	8 9
1	3 4 4
2	2 3 7 8
3	1 1 4
4	2 3
5	1

3 | 1 means 31 years

.... years **(2 marks)**

3 Sophie collected some information about the heights, in cm, of some shrubs.

1	7 8
2	2 5 6 7
3	3 4 8 8 9
4	2 3 5
5	5 6 7
6	1 7

3 | 3 means 33 cm

(a) How many shrubs did Sophie measure?

.... **(1 mark)**

(b) Work out the range.

.... cm **(1 mark)**

(c) Work out the median.

.... cm **(1 mark)**

(d) Work out the interquartile range

.... cm **(2 marks)**

Line graphs

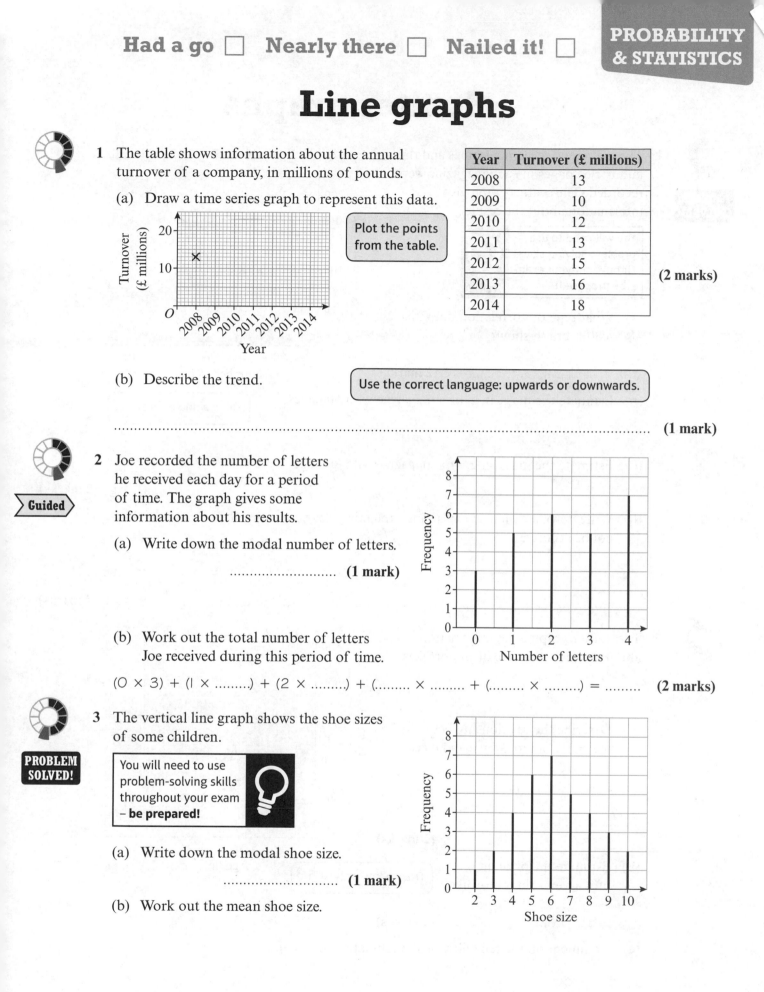

1   The table shows information about the annual turnover of a company, in millions of pounds.

Year	Turnover (£ millions)
2008	13
2009	10
2010	12
2011	13
2012	15
2013	16
2014	18

(a) Draw a time series graph to represent this data.

> Plot the points from the table.

(2 marks)

Turnover (£ millions) / Year

(b) Describe the trend.

> Use the correct language: upwards or downwards.

.... **(1 mark)**

Guided

2   Joe recorded the number of letters he received each day for a period of time. The graph gives some information about his results.

(a) Write down the modal number of letters.

.... **(1 mark)**

Frequency / Number of letters

(b) Work out the total number of letters Joe received during this period of time.

$(0 \times 3) + (1 \times ....) + (2 \times ....) + (........ \times .... + (........ \times ....) = ...$ **(2 marks)**

PROBLEM SOLVED!

3   The vertical line graph shows the shoe sizes of some children.

> You will need to use problem-solving skills throughout your exam – **be prepared!**

Frequency / Shoe size

(a) Write down the modal shoe size.

.... **(1 mark)**

(b) Work out the mean shoe size.

.... **(3 marks)**

Scatter graphs

PROBLEM SOLVED!

1 The masses of seven magazines and the number of pages in each magazine were recorded. The scatter graph gives information about the results.

> You will need to use problem-solving skills throughout your exam – **be prepared!**

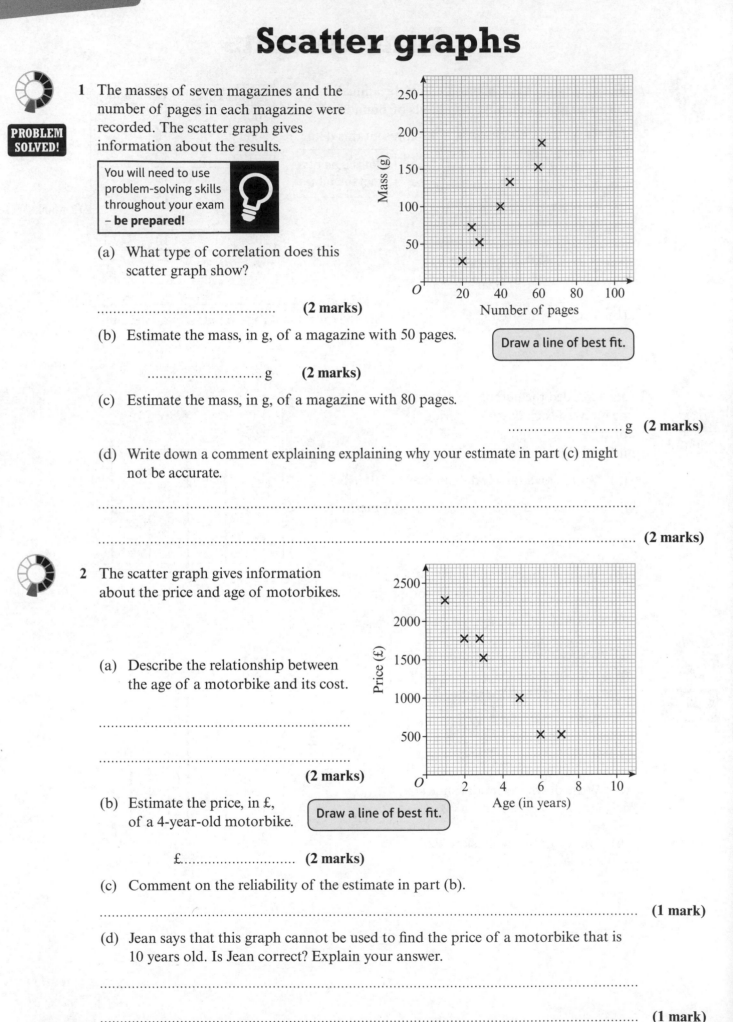

(a) What type of correlation does this scatter graph show?

.... **(2 marks)**

(b) Estimate the mass, in g, of a magazine with 50 pages.

> Draw a line of best fit.

.... g   **(2 marks)**

(c) Estimate the mass, in g, of a magazine with 80 pages.

.... g   **(2 marks)**

(d) Write down a comment explaining explaining why your estimate in part (c) might not be accurate.

....

....   **(2 marks)**

2 The scatter graph gives information about the price and age of motorbikes.

(a) Describe the relationship between the age of a motorbike and its cost.

....

....   **(2 marks)**

(b) Estimate the price, in £, of a 4-year-old motorbike.

> Draw a line of best fit.

£...........................   **(2 marks)**

(c) Comment on the reliability of the estimate in part (b).

....   **(1 mark)**

(d) Jean says that this graph cannot be used to find the price of a motorbike that is 10 years old. Is Jean correct? Explain your answer.

....

....   **(1 mark)**

Sampling

1 Tony is conducting a survey for a magazine on the spending habits of people at a local shopping centre. He plans to survey the first 9 people in a queue at one shop on one Sunday morning for a sample.

(a) Write down one advantage of taking a sample.

.... **(1 mark)**

(b) Comment on the reliability of this sample.

....

.... **(1 mark)**

(c) Write down one way of reducing bias in this sample.

....

.... **(1 mark)**

2 An experiment is carried out by flying paper aeroplanes. The scatter graph shows some information about the distances flown, in m, and the wingspans, in cm. A line of best fit has been drawn.

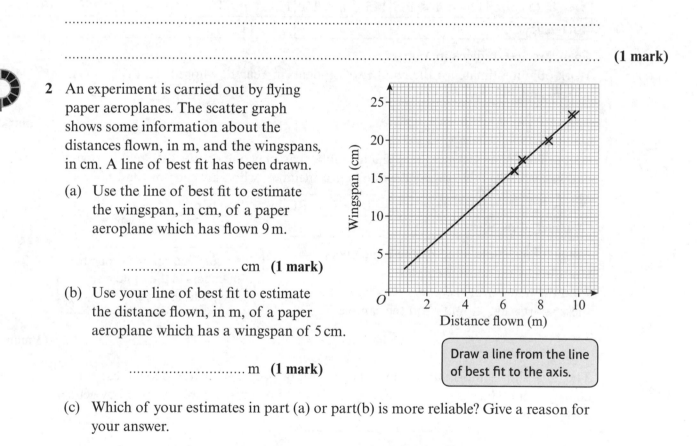

(a) Use the line of best fit to estimate the wingspan, in cm, of a paper aeroplane which has flown 9 m.

.... cm **(1 mark)**

(b) Use your line of best fit to estimate the distance flown, in m, of a paper aeroplane which has a wingspan of 5 cm.

.... m **(1 mark)**

> Draw a line from the line of best fit to the axis.

(c) Which of your estimates in part (a) or part (b) is more reliable? Give a reason for your answer.

....

.... **(2 marks)**

(d) Write down one way you could improve this experiment to increase the accuracy of your estimates.

....

.... **(1 mark)**

Stratified sampling

1   A school has 750 students. Each student studies one of Latin, Spanish, French and German. The table shows the number of students who study each of these languages.

Guided

Language	Latin	Spanish	French	German
Number of students	170	146	220	214

Peter takes a sample of 50 of these students stratified by the language studied.
Find the number of students in his sample who study German.

$$\frac{214}{\text{.........}} \times \text{.............................} = \text{.............................}$$

(2 marks)

2   The grouped frequency table shows information about the heights, in centimetres, of 30 students, chosen at random from Year 12.

Height (h cm)	$120 < h \leqslant 135$	$135 < h \leqslant 150$	$150 < h \leqslant 165$	$165 < h \leqslant 180$	$180 < h \leqslant 195$
Frequency	4	5	11	7	3

There are 360 students in Year 12.
Work out an estimate for the number of students in Year 12 whose height is between 150 cm and 165 cm.

.... **(3 marks)**

3   A youth club has 450 members. Each member can play one of football, tennis, rugby and squash. The table shows the number of members who play each of these sports.

Sport	Football	Tennis	Rugby	Squash
Number of members	95	68	151	136

Bill takes a sample of 65 of these members, stratified by the sport they play.
Find the number of members playing each of these sports that should be in the sample.

> Make sure you count the number of members in the sample so that it adds up to 65.

.... **(3 marks)**

4   The table shows information about the number of people who attended a local charity event. Gary is going to take a sample of 55 of these people stratified by gender and by age.

		Age		
		Under 19	**19 to 39**	**Over 39**
Gender	**Male**	136	183	85
	Female	158	200	138

Calculate the number of males aged over 39 that should be in his sample.

.... **(3 marks)**

Capture–recapture

1 Alice wants to estimate the number of frogs in a lake. She catches a sample of 70 frogs, marks them and puts them back in the lake.
Later that day, in a second sample of 60 frogs, she finds that 8 of them are marked.
Work out an estimate for the number of frogs in the lake.

Using $N = \dfrac{Mn}{m}$

$$\frac{Mn}{m} = \frac{\text{.......} \times \text{.......}}{8}$$

$$= \text{.............................}$$

(3 marks)

2 Bart wants to find an estimate for the number of ants in a colony. He catches 75 ants from the colony and marks each one with a dye. He then returns the ants to the colony.
A week later, Bart catches another 80 ants. Eight of these ants are marked with the dye.
Work out an estimate for the number of ants in the colony.

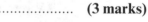 **(3 marks)**

3 Carl has a box containing a large number of beads.
He wants to find an estimate for the number of beads in the box.
Carl takes a sample of 40 beads from the box.
He marks each bead with a black tick. He then puts the beads back in the box and shakes the box.
He now takes another sample of 50 beads from the box. Four of these beads have been marked with a black tick.
Work out an estimate for the total number of beads in the box.

You will need to use problem-solving skills throughout your exam – **be prepared!**

PROBLEM SOLVED!

.... **(3 marks)**

Cumulative frequency

1 The cumulative frequency diagram gives information about the masses in grams of a number of letters.

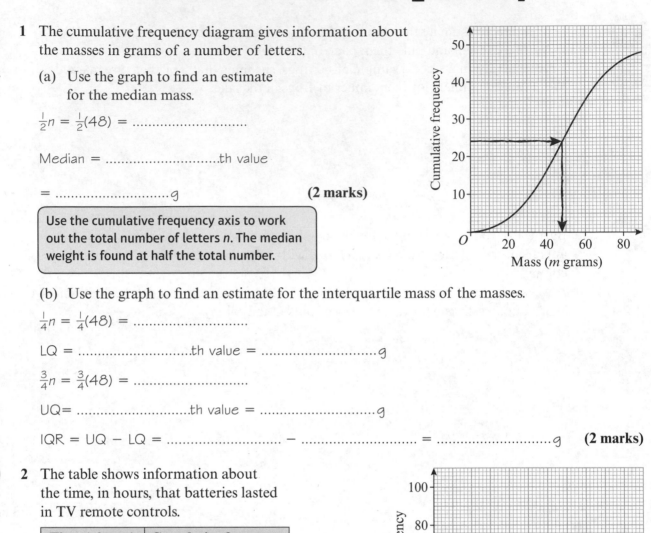

Guided

(a) Use the graph to find an estimate for the median mass.

$\frac{1}{2}n = \frac{1}{2}(48) = $ ....

Median = ...th value

= ....g   **(2 marks)**

> Use the cumulative frequency axis to work out the total number of letters n. The median weight is found at half the total number.

(b) Use the graph to find an estimate for the interquartile mass of the masses.

$\frac{1}{4}n = \frac{1}{4}(48) = $ ....

LQ = ...th value = ...g

$\frac{3}{4}n = \frac{3}{4}(48) = $ ....

UQ = ...th value = ...g

IQR = UQ − LQ = ... − .... = ...g   **(2 marks)**

2 The table shows information about the time, in hours, that batteries lasted in TV remote controls.

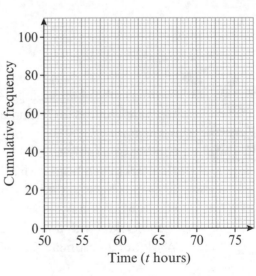

Time (t hours)	Cumulative frequency
$50 \leqslant t < 55$	11
$50 \leqslant t < 60$	32
$50 \leqslant t < 65$	68
$50 \leqslant t < 70$	93
$50 \leqslant t < 75$	100

(a) On the grid, draw a cumulative frequency graph to show this information.   **(2 marks)**

(b) Use the graph to find an estimate for the median.

.... hours   **(2 marks)**

(c) Use the graph to find an estimate for the interquartile range.

.... hours   **(2 marks)**

(d) Tom says at least 20% of the batteries lasted more than 68 hours. Is Tom correct? You must give a reason for your answer.

....   **(2 marks)**

Box plots

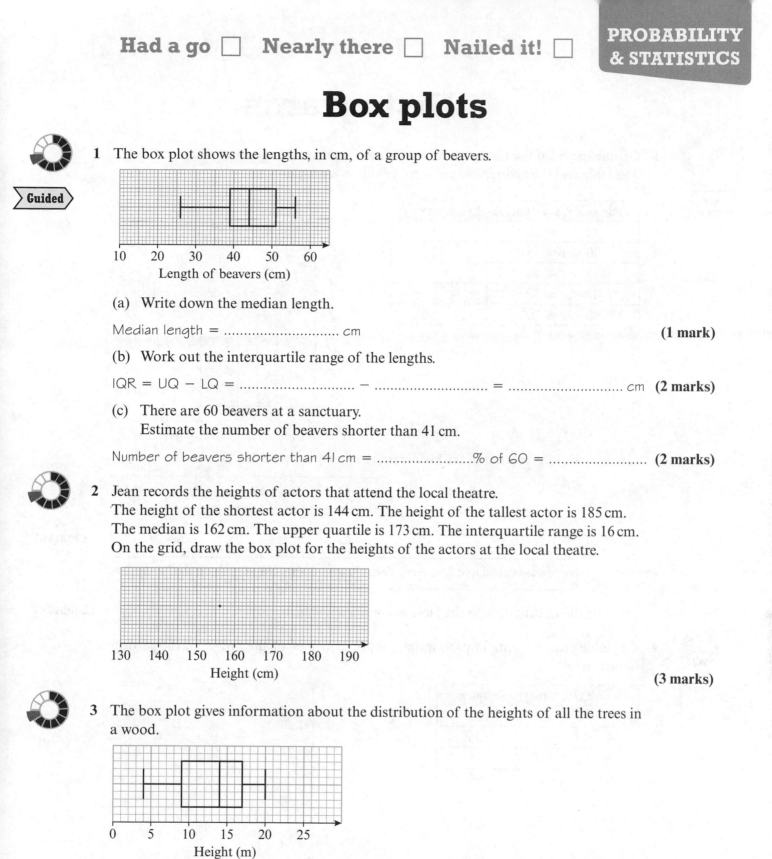

1   The box plot shows the lengths, in cm, of a group of beavers.

> **Guided** >

Length of beavers (cm)

(a) Write down the median length.

Median length = ... cm   **(1 mark)**

(b) Work out the interquartile range of the lengths.

IQR = UQ − LQ = ... − .... = ... cm   **(2 marks)**

(c) There are 60 beavers at a sanctuary.
Estimate the number of beavers shorter than 41 cm.

Number of beavers shorter than 41 cm = ...% of 60 = ...   **(2 marks)**

2   Jean records the heights of actors that attend the local theatre.
The height of the shortest actor is 144 cm. The height of the tallest actor is 185 cm.
The median is 162 cm. The upper quartile is 173 cm. The interquartile range is 16 cm.
On the grid, draw the box plot for the heights of the actors at the local theatre.

Height (cm)

                                                                            (3 marks)

3   The box plot gives information about the distribution of the heights of all the trees in
a wood.

Height (m)

(a) What percentage of trees are taller than 9 m?

............................% **(1 mark)**

(b) There are 240 trees in the wood. Estimate how many trees are between 9 m and 14 m

.... **(2 marks)**

Histograms

1 George recorded the time it took him to travel to work.
The table and the histogram give some of this information.

Time taken (t minutes)	Frequency (f)
$20 < t \leqslant 30$	7
$30 < t \leqslant 35$	12
$35 < t \leqslant 40$	
$40 < t \leqslant 50$	
$50 < t \leqslant 70$	8

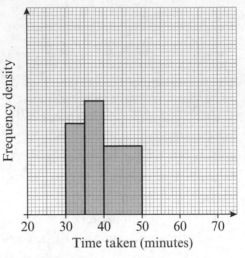

(a) Use the information in the histogram to complete the table.

> Frequency density, FD = frequency ÷ class width

FD of $30 < t \leqslant 35$ = ... ÷ .... = ... **(2 marks)**

> Use your calculated frequency densities to number the FD axis on the histogram.

(b) Use the information in the table to complete the histogram.   **(2 marks)**

2 The table gives information about the lengths, in metres, of the hedges of the gardens in one street.

Length (l metres)	Frequency
$0 < l \leqslant 40$	25
$40 < l \leqslant 60$	36
$60 < l \leqslant 80$	14
$80 < l \leqslant 90$	5

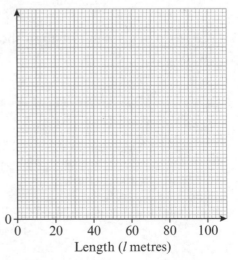

(a) Draw a histogram to represent the information in the table.   **(3 marks)**

(b) Estimate the number of hedges between 48 m and 84 m in length.

.... **(2 marks)**

Frequency polygons

1 Jay asked some people how many minutes they each took to walk to the library.
The table shows some information about his results.

Guided

Time taken (*t* minutes)	Frequency
$0 < t \leqslant 10$	3
$10 < t \leqslant 20$	8
$20 < t \leqslant 30$	10
$30 < t \leqslant 40$	4

(a) On the grid, draw a frequency polygon to represent the information in the table.

> Add an extra column to the table for the midpoints of each class interval. Plot the points for each midpoint.

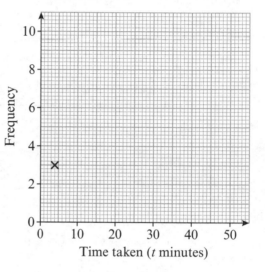

> Join the 4 points with 3 straight lines.

1st point at (5, 3)

2nd point at (15, ...)   **(2 marks)**

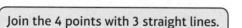

(b) Write down the modal class.

.... **(1 mark)**

2 Kerry recorded the masses, in kg, of 70 ornaments.

Mass (*m* kg)	Frequency
$2.5 < m \leqslant 3.5$	6
$3.5 < m \leqslant 4.5$	10
$4.5 < m \leqslant 5.5$	15
$5.5 < m \leqslant 6.5$	22
$6.5 < m \leqslant 7.5$	17

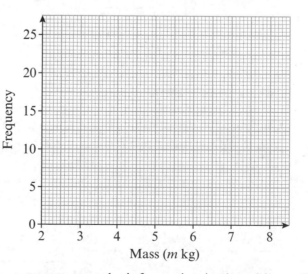

(a) On the grid, draw a frequency polygon to represent the information in the table.   **(2 marks)**

(b) Write down the class interval which contains the median.

.... **(1 mark)**

Comparing data

1   The exam marks of classes 11A and 11B are shown in the back-to-back stem-and-leaf diagram.

```
        11A           11B
    5 4 1 │ 5 │
  8 6 3 2 │ 6 │ 2 3 4
  8 7 5 4 │ 7 │ 1 3 6 7
        2 │ 8 │ 5 6 8
          │ 9 │ 3
```

Key: 1 | 5 represents 51 marks   Key: 6 | 2 represents 62 marks

Compare the results of class 11A with the results of class 11B.

> Work out the median and the range and then use these values to compare the data.

The median for class IIA is .... The range for class IIA

is .... − .... = ...

....

....

.... **(2 marks)**

2   These box plots give information about the times taken by Anjali and Carol to each complete some puzzles.

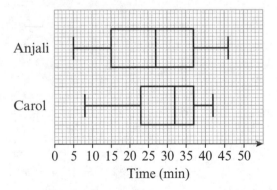

Compare the distributions of the times taken for Anjali and Carol to complete the puzzles. Write down two comparisons.

....

....

.... **(2 marks)**

> Always compare median and range or interquartile range.

Probability

1 A box contains cartons of orange juice, apple juice and mango juice.
The table shows each of the probabilities that a carton of juice taken at
random from the box will be orange or apple.

Guided

Carton of juice	Orange	Apple	Mango
Probability	0.3	0.4	

> The probabilities must add up to 1.

A carton is to be taken at random from the box.
Work out the probability that the carton

(a) will be an orange juice or an apple juice

0.3 + ... = ...   **(2 marks)**

(b) will be a mango juice.

1 − (............... + ...) = ...   **(2 marks)**

2 A bag contains red, green, white and blue counters. The table shows each of the
probabilities that a counter taken at random will be red or green or white.

Guided

Colour	Red	Green	White	Blue
Probability	0.35	0.28	0.16	

A counter is to be taken at random from the bag.
Work out the probability that the counter will be blue.

1 − (............... + ... + ...) = ...   **(2 marks)**

3 A spinner can land on A, B, C or D. The table shows information that the spinner
will land on each letter B or C or D.

Letter	A	B	C	D
Probability		0.26	0.36	0.17

The spinner is spun once. Work out the probability that the spinner will land on

(a) B or C   (b) A.

.... **(2 marks)**   .... **(2 marks)**

4 Four athletes Andy, Ben, Carl and Daljit take
part in a race. The table shows the probabilities
of Andy or Ben winning the race.

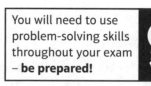

You will need to use
problem-solving skills
throughout your exam
– **be prepared!**

PROBLEM SOLVED!

Athlete	Andy	Ben	Carl	Daljit
Probability	0.3	0.38		

The probability that Carl will win is 3 times the probability that Daljit will win.
Work out the probability that the race will be won by

(a) Andy or Ben   (b) Daljit.

.... **(2 marks)**   .... **(2 marks)**

Relative frequency

Guided

1 The table shows information about the number of orders received each month for six months by an internet company.

Month	Jan	Feb	Mar	Apr	May	Jun
Number of orders	28	63	49	61	53	48

An order is chosen at random.
Work out the probability that the order was received in

(a) May

> First work out the total number of orders.

53
‾‾‾‾‾

....   **(2 marks)**

(b) Jan or Feb or Mar

> Add up the numbers for Jan, Feb and Mar.

28 + ... + ... = ...

....
‾‾‾‾‾‾‾   **(2 marks)**
....

Guided

2 The table shows the total scores when Ethan throws three darts 50 times.

Score	1–30	31–60	61–90	91–120	121–150	151–180
Frequency	14	10	9	8	6	3

He throws another three darts. Estimate the probability that he scores

(a) between 31 and 60

.... out of 50 = $\dfrac{.....................}{50}$ **(1 mark)**

(b) more than 90

.... **(2 marks)**

(c) 120 or less.

.... **(2 marks)**

3 A garage keeps records of the cost of repairs it makes to vans. The table gives information about the costs of all repairs which were less than £500 in one month.

Cost (£C)	Frequency
$0 < C \leqslant 100$	20
$100 < C \leqslant 200$	39
$200 < C \leqslant 300$	72
$300 < C \leqslant 400$	33
$400 < C \leqslant 500$	38

(a) Amy needs to repair her van. Estimate the probability that the cost of her repair is more than £200.

.... **(2 marks)**

(b) Comment on the accuracy of your estimate.

.... **(1 mark)**

Venn diagrams

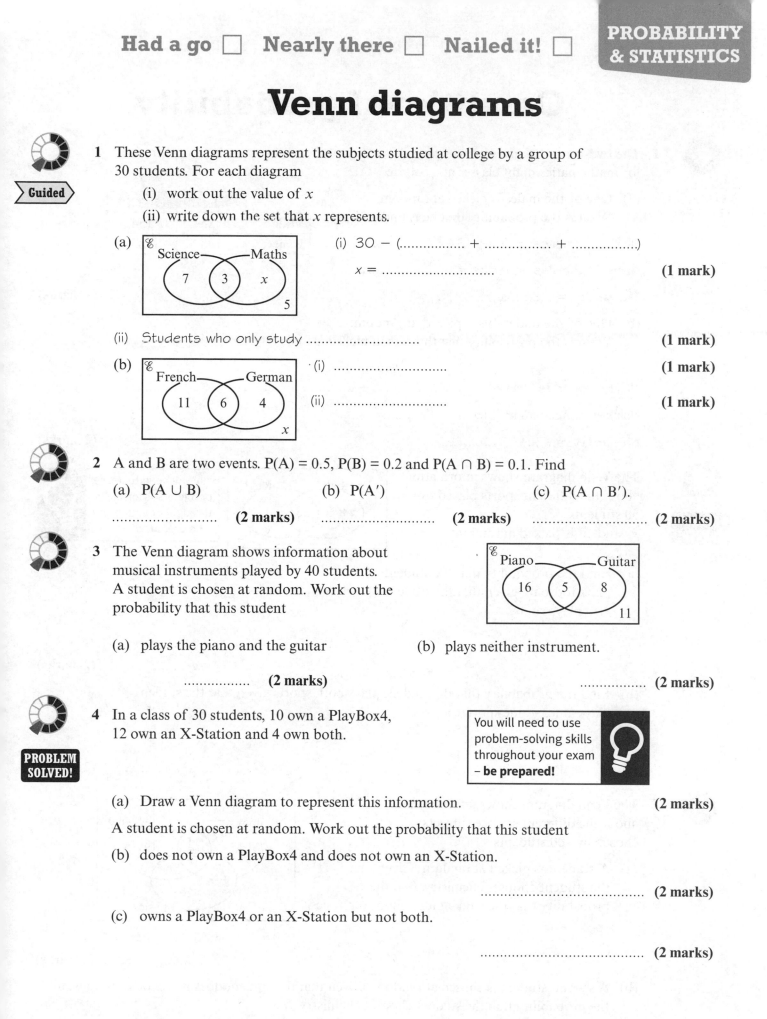

1   These Venn diagrams represent the subjects studied at college by a group of 30 students. For each diagram

   (i)   work out the value of x

   (ii) write down the set that x represents.

(a)   Science — Maths   7   3 x 5

(i)   30 − (............... + ... + ...)

    x = ...   **(1 mark)**

(ii)   Students who only study ....   **(1 mark)**

(b)   French — German   11   6   4 x

(i)   ....   **(1 mark)**

(ii)   ....   **(1 mark)**

2   A and B are two events. P(A) = 0.5, P(B) = 0.2 and P(A ∩ B) = 0.1. Find

(a)   P(A ∪ B)

(b)   P(A′)

(c)   P(A ∩ B′).

.... **(2 marks)**   .... **(2 marks)**   .... **(2 marks)**

3   The Venn diagram shows information about musical instruments played by 40 students. A student is chosen at random. Work out the probability that this student

Piano — Guitar   16   5   8   11

(a)   plays the piano and the guitar

(b)   plays neither instrument.

.... **(2 marks)**   .... **(2 marks)**

4   In a class of 30 students, 10 own a PlayBox4, 12 own an X-Station and 4 own both.

You will need to use problem-solving skills throughout your exam – **be prepared!**

PROBLEM SOLVED!

(a)   Draw a Venn diagram to represent this information.   **(2 marks)**

A student is chosen at random. Work out the probability that this student

(b)   does not own a PlayBox4 and does not own an X-Station.

.... **(2 marks)**

(c)   owns a PlayBox4 or an X-Station but not both.

.... **(2 marks)**

Conditional probability

1 The two-way table gives information about the number of people who have enrolled for mathematics night classes at a college.

(a) One of the males is picked at random.
   What is the probability that he is under 19?

	Under 19	Over 19
Male	85	134
Female	198	83

Number of males = 85 + 134 = ...

Number of males under 19 = ...

Probability = ...

(2 marks)

(b) One of the under-19s is picked at random.
   What is the probability that the person picked is male?

Number of under-19s = ... + ... = ...

Number of males under 19 = ...

Probability = ...

(2 marks)

2 The Venn diagram shows information about the different sports played by 50 students.
A student is picked at random.

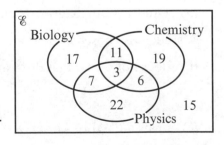

You will need to use problem-solving skills throughout your exam – **be prepared!**

PROBLEM SOLVED!

(a) Find the probability that the student plays football, given that the student also plays rugby.

.... **(2 marks)**

(b) Find the probability that the student plays both sports given that the student plays at least one sport.

.... **(2 marks)**

3 The Venn diagram shows information about the different science subjects chosen by 100 students.

(a) A student is picked at random. Given that the student chooses chemistry, find the probability that the student also chooses physics.

.... **(2 marks)**

(b) A second student is chosen at random. Given that the student does not choose biology, find the probability that the student chooses chemistry.

.... **(3 marks)**

Tree diagrams

1 A jar contains 4 yellow sweets and 6 blue sweets.
A sweet is picked at random and not replaced.
A second sweet is then picked.

> **Guided**

(a) Complete the tree diagram for this information.

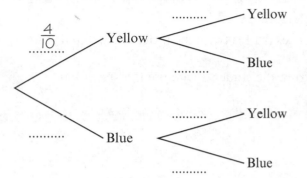

(2 marks)

Find the probability that

(b) the second sweet is yellow given that the first sweet is blue

$$\frac{\text{.........}}{9}$$

(1 mark)

(c) both sweets are yellow

$$\frac{\text{.........}}{10} \times \frac{\text{.........}}{9} = \frac{\text{.........}}{\text{.........}} = \frac{\text{.........}}{\text{.........}}$$

(2 marks)

(d) the sweets are of different colours.

$$\left(\frac{\text{.........}}{10} \times \frac{\text{.........}}{9}\right) + \left(\frac{\text{.........}}{10} \times \frac{\text{.........}}{9}\right) = \frac{\text{.........}}{\text{.........}} + \frac{\text{.........}}{\text{.........}} = \frac{\text{.........}}{\text{.........}} = \frac{\text{.........}}{\text{.........}}$$

(3 marks)

2 The probability that a certain disease occurs in a population is 0.15.
If the patient has the disease, the probability that a screening procedure produces a positive result is 0.7.
If the patient does not have the disease, there is still a 0.1 chance that the test will give a positive result.
Find the probability that a randomly selected individual

(a) does not have the disease but gives a positive result in the screening test

> Draw a tree diagram.

.... (3 marks)

(b) gives a positive result in the test.

.... (2 marks)

Problem-solving practice 1

1 A school snack bar offers a choice of four snacks. The four snacks are burgers, wraps, fruit and salad. Students can choose one of these four snacks. The table shows the probability that a student will choose a burger or a wrap.

Snack	Burger	Wrap	Fruit	Salad
Probability	0.25	0.15		

The probability that a student chooses fruit is twice the probability that a student of chooses salad.

One student is picked at random from the students who use the snack bar.

(a) Work out the probability that the student

   (i) does not choose a burger

.... **(2 marks)**

   (ii) chooses salad.

.... **(2 marks)**

(b) 200 students used the snack bar on Tuesday.
Work out an estimate for the number of students who chose a wrap.

.... **(2 marks)**

2 The heights, in cm, of some plants were measured in Park A and in Park B. The information is shown in the back-to-back stem-and-leaf diagram.

Compare the heights of plants in Park A with the heights of plants in Park B.

```
   Park A          Park B
              1 | 0  3  4
      3  3  1 | 2 | 1  2  5  7
   6  5  2  0 | 3 | 3  4  6  7
      7  5  4 | 4 | 2
            2 | 5 |
```

Key: 1 | 2 represents 21 cm   Key: 1 | 0 represents 10 cm

....

.... **(2 marks)**

3 In a class of 25 students, 8 study Latin, 10 study Mandarin and 3 study both.
A student is picked at random. Find the probability that the student

(a) does not study Latin and does study Mandarin

.... **(3 marks)**

(b) studies Latin or Mandarin but not both.

.... **(2 marks)**

Problem-solving practice 2

4 The histogram shows information about the salaries of a sample of people.

Work out the proportion of people in the sample who have a salary between £10 000 and £35 000.

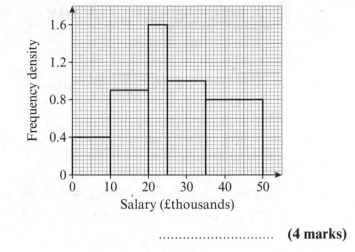

Salary (£thousands)

.... **(4 marks)**

5 There are n counters in a jar.
7 of the counters are red. The rest of the counters are green.
Keiran takes a counter at random from the jar and places it on the table.
He then takes another counter at random from the jar and places it on the table.
The probability that Keiran takes two red counters is $\frac{1}{5}$.
Work out the number of counters in the jar.

.... **(5 marks)**

6 80 students each study one of three languages.
The languages are French, German and Spanish.
There are 39 females who study a language. 15 females study French.
17 of the males study German. 8 of the 21 students who study Spanish are male.
One of these students is picked at random.

(a) Work out the probability that the student picked studies French.

.... **(4 marks)**

(b) Given that the student studies German, find the probability that the student is male.

.... **(2 marks)**

Paper 1

Practice exam paper

Higher Tier
Time: 1 hour 30 minutes
Calculators must not be used
Diagrams are **NOT** accurately drawn,
unless otherwise indicated.
You must **show all your working out**.

1   Here are the ingredients needed to make 6 hotcakes.

   Hotcakes
   Makes 6 hotcakes
   50 g sugar
   200 g butter
   200 g flour
   10 ml milk

   Peter makes some hotcakes.
   He uses 15 ml milk.

   (a)   How many hotcakes does Peter make?

   .... **(1 mark)**

   Asha has
   600 g sugar
   1200 g butter
   1350 g flour
   430 ml milk

   (b)   Work out the greatest number of hotcakes Asha can make.

   .... **(2 marks)**

2   Here are the speeds, in miles per hour, of 16 cars.

   41   62   53   59   46   45   43   39
   64   53   54   56   52   49   65   58

   Draw an ordered stem-and-leaf diagram for these speeds.

   (3 marks)

3   *ABC* is parallel to *DEFG*. *BE* = *EF*. Angle *ABE* = 42°.

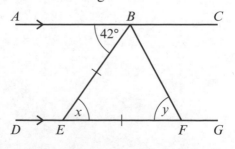

(a) (i) Work out the size of the angle marked x.

x = ...

(ii) Give a reason for your answer.

.... **(2 marks)**

(b) Work out the size of the angle marked y.

y = ... **(1 mark)**

4 Pavan comes back from the USA with some US dollars ($) which he wants to change into pounds (£). He uses the exchange rate £1 = $1.50

(a) On the grid, draw a conversion graph Pavan can use to change between pounds and dollars.

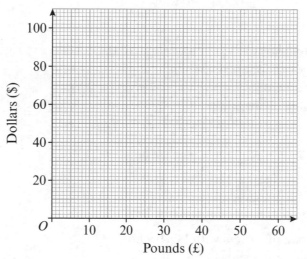

(2 marks)

Pavan changes $1000 into pounds.

(b) Use your graph to work out how many pounds he gets for $1000

£................ **(2 marks)**

5 This is some information about a class.
- There are 40 students in the class.
- 16 of the students study Latin.
- 19 of the students study Spanish.
- 7 of the students study both Latin and Spanish.

(a) Draw a Venn diagram to represent this information.

(4 marks)

(b) Show that 30% of the students do not study Latin or Spanish.

(2 marks)

131

6 Expand and simplify $(x + 5)(x - 2)$

.... **(2 marks)**

7 Here is a six-sided shape.

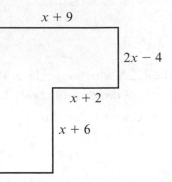

All the measurements are in centimetres. All the corners are right angles.
The perimeter of the shape is 94 cm.
Work out the value of x.
You must show your working.

(5 marks)

8 Amy shares a bag of sweets with her friends.
She gives Beth $\frac{2}{5}$ of the sweets.
She gives Carl $\frac{3}{10}$ of the sweets.
She has 18 sweets left.
How many sweets does Amy give to Beth?

.... **(4 marks)**

9 A, C and B are three places on a map.

ACB is a straight line.

Construct the perpendicular to the line AB at point C.

You must leave all your construction lines.

A ———————×——————— B
      C

(2 marks)

10 A force of 1800 newtons is applied to a square piece of metal plate measuring 30 cm × 30 cm.

Using the formula pressure = $\dfrac{\text{force}}{\text{area}}$, work out the pressure exerted on the metal plate, in N/m².

.... N/m² **(2 marks)**

11 A circular chopping board has a radius of 80 mm.

(a) Work out the area of the chopping board, in mm².
Leave your answer in terms of π.

.... mm² **(2 marks)**

(b) The volume of the chopping board is $115\,200\,\pi\,\text{mm}^3$.
Work out the thickness of the chopping board.

.... mm **(2 marks)**

12 (a) (i) Write 50 000 in standard form.

....

(ii) Write 9.6×10^{-5} as an ordinary number.

.... **(2 marks)**

(b) Work out the value of $(5 \times 10^4) \times (3 \times 10^6)$.
Give your answer in standard form.

.... **(2 marks)**

13 Solve the simultaneous equations
$5x + 2y = 11$
$4x - 3y = 18$

.... **(4 marks)**

14 The speed, s, of a particle is inversely proportional to the time, t, taken.
When $s = 12$, $t = 4$

(a) Find a formula for s in terms of t.

.... **(3 marks)**

(b) Hence, or otherwise, calculate the value of s when $t = 3$

$s =$ .... **(1 mark)**

15 The diagram shows the points B, C and D on a circumference of a circle with centre O.

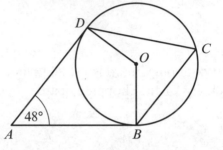

AB and AD are tangents to the circle. Angle $DAB = 48°$
Work out the size of angle BCD.
Give a reason for each stage in your working.

...................................° **(4 marks)**

16 Solve $x^2 - 8x > 20$

.... **(4 marks)**

17 A solid hemisphere has a radius of x.
A solid cone has a **diameter** of $3x$ and a perpendicular height of y.
The hemisphere and the cone have equal volumes.

Show that $y = \dfrac{8}{9}x$

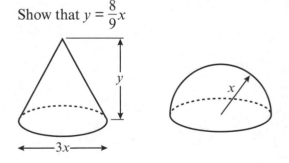

Volume of cone $= \dfrac{1}{3}\pi r^2 h$

Volume of sphere $= \dfrac{4}{3}\pi r^3$

(3 marks)

134

18 (a) Rationalise the denominator of $\dfrac{12}{\sqrt{3}}$

.... **(2 marks)**

(b) Expand $(2 + \sqrt{5})(3 + \sqrt{5})$

Give your answer in the form $a + b\sqrt{5}$ where a and b are integers.

.... **(2 marks)**

19 The table gives some information about the speeds, in km/h, of 120 motorbikes.

Speed(s km/h)	Frequency
$60 < s \leqslant 80$	32
$80 < s \leqslant 90$	42
$90 < s \leqslant 95$	26
$95 < s \leqslant 100$	20

(a) On the grid, draw a histogram for the information in the table.

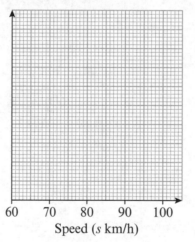

Speed (s km/h)

(3 marks)

(b) Work out an estimate for the number of motorbikes with a speed of less than 75 km/h.

.... **(2 marks)**

20 Simplify fully $\dfrac{x^2 - 2x - 15}{2x^2 - 9x - 5}$

.... **(3 marks)**

21 Write $\dfrac{2}{x + 4} + \dfrac{3}{x - 2}$ as a single fraction in its simplest form.

.... **(3 marks)**

22 The graph of $y = f(x)$ is shown on each of the grids.

(a) On this grid, sketch the graph of $y = f(x + 2)$

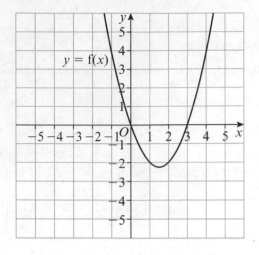

(2 marks)

(b) On this grid, sketch the graph of $y = -f(x)$

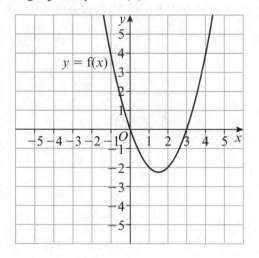

(2 marks)

TOTAL FOR PAPER = 80 MARKS

Practice exam papers for Paper 2 and Paper 3 are available to download free from the Pearson website. Scan this QR code or visit http://activetea.ch/1GYGg8W.

Answers

NUMBER

1. Factors and primes
1. (a) (i) $2 \times 3^2 \times 5$
       (ii) $2 \times 3 \times 5 \times 7$
   (b) 30   (c) 630
2. 25, 50
3. $m = 2, n = 5$
4. 3 packs of hats, 2 packs of scarves

2. Indices 1
1. (a) 4^2 (b) 5
2. (a) 16   (b) 8   (c) 8
   (d) 4   (e) 3   (f) -4
3. (a) 5^9 (b) 5^3 (c) 5^4
4. (a) 3^3 (b) 3^2 (c) 3^8 (d) 3^6
5. (a) 9   (b) 4
6. 8
7. $2^3 = 8$ not 6
8. $x = 7$ and $y = 5$

3. Indices 2
1. (a) $\frac{1}{8}$ (b) $\frac{1}{3}$ (c) $\frac{1}{49}$ (d) $\frac{1}{2}$
2. (a) $\frac{1}{3}$ (b) 4   (c) $\frac{5}{3}$ (d) $\frac{7}{9}$
3. (a) $\frac{4}{9}$ (b) $\frac{64}{27}$ (c) $\frac{16}{25}$ (d) $\frac{1}{125}$
4. (a) $\frac{9}{16}$ (b) 27   (c) $\frac{25}{36}$ (d) $\frac{125}{27}$
5. (a) 5   (b) 2   (c) 4   (d) 3
6. (a) 64   (b) 8   (c) 125   (d) 9
7. $8^{\frac{1}{3}} = 2$ so $8^{\frac{2}{3}} = 2^2 = 4$
   $16^{\frac{1}{2}} = 4$
8. (a) xy (b) y^2

4. Calculator skills 1
1. (a) 3380   (b) 9.75   (c) 6.87
2. (a) 216   (b) 17.9
3. 1.751 592 357
4. (a) 8.5625   (b) 9
5. (a) 2.27×10^4 (b) 2.22×10^9
6. (a) 2.47   (b) 5.13
7. 3.00×10^9

5. Fractions
1. (a) $6\frac{11}{20}$ (b) $2\frac{1}{10}$
2. (a) $3\frac{5}{6}$ (b) $3\frac{1}{3}$
3. (a) 9   (b) $3\frac{9}{13}$
4. £80
5. $\frac{7}{24}$
6. 16
7. $5\frac{11}{12}$ hours
8. $16\frac{2}{3} \, \text{m}^2$

6. Decimals
1. $0.3, \frac{1}{3}, 0.35, \frac{18}{50}$
2. 0.15
3. $2 \times 3 \times 5$
4. (a) $2^3 \times 5$ terminating   (b) 2^5 terminating
   (c) 3×13 recurring   (d) $2 \times 3 \times 7$ recurring
5. 0.181 818 181 8
6. (a) 0.275   (b) 0.24   (c) 0.367
7. No, because $12 \div 5 = 2.4$ not 2.375
8. (a) 11 7300   (b) 1.173   (c) 8500

7. Estimation
1. (a) 14 000   (b) 6   (c) 125 000
2. 125
3. 7200
4. 17 500
5. 750
6. (a) 432 cm²
   (b) Underestimate, because the original values have been rounded down
7. (a) 1500 m²
   (b) Overestimate, because the original values have been rounded up

8. Standard form
1. (a) 4.5×10^4 (b) 0.000 034   (c) 2.8×10^7
2. (a) 5.67×10^5 (b) 5.67×10^{-5} (c) 5.67×10^{10}
3. (a) 2.05×10^8 (b) 7.5×10^7
4. (a) 1.8×10^4 (b) 2×10^{20}
5. (a) 7.01×10^4 (b) 7.52×10^5
6. 1.44×10^8 km

9. Recurring decimals
1. $x = 0.777\,77$
   $10x = 7.777\,77$
   Subtracting,
   $9x = 7$ so $x = \frac{7}{9}$
2. $x = 0.424\,242\,42$
   $100x = 42.424\,242$
   Subtracting,
   $99x = 42$ so $x = \frac{42}{99} = \frac{14}{33}$
3. (a) $x = 0.818\,181\,81$
       $100x = 81.818\,181$
       Subtracting,
       $99x = 81$ so $x = \frac{81}{99} = \frac{9}{11}$
   (b) $\frac{53}{110}$
4. $\frac{611}{990}$
5. $\frac{52\,319}{9990}$
6. $x = 6.432\,222\,22$
   $100x = 643.222\,22$
   $1000x = 6432.222\,22$
   Subtracting,
   $900x = 5789$ so $x = \frac{5789}{900}$
7. $y = 0.0x0x0x0x$
   $100y = 00x.0x0x0x$
   Subtracting,
   $99y = x$ so $y = \frac{x}{99}$

10. Upper and lower bounds
1. (a) 19.5 kg   (b) 20.5 kg
2. (a) 52.35 cm   (b) 52.25 cm
3. LB = 1360 J; UB = 1420 J
4. 11.37 to 11.42 g/cm³
5. 64.1 m

11. Accuracy and error
1. 33.35 cm ≤ length of rod < 33.45 cm
2. 1.745 litres ≤ volume < 1.755 litres
3. 6.5 cm (LB = 6.46..., UB = 6.49... both round to 6.5 to 2 s.f.)
4. 54 is the safest maximum
5. Yes. Ravina's handspan could be as small as 145 mm whereas Anjali's could be as large as 148.5 mm
6. (a) 3.079 and 3.128   (b) 3.1 m/s

12. Surds 1
1. (a) $2\sqrt{3}$ (b) $2\sqrt{5}$ (c) $4\sqrt{3}$
   (d) 9   (e) $4\sqrt{2}$ (f) $7\sqrt{7}$
2. (a) $\frac{3\sqrt{5}}{5}$ (b) $\frac{\sqrt{6}}{3}$ (c) $\frac{\sqrt{2}}{2}$
   (d) $3\sqrt{7}$ (e) $\frac{3 + \sqrt{3}}{6}$ (f) $3 + 5\sqrt{2}$
3. (a) 2   (b) 10
4. $2\sqrt{3}$

$5 \quad \dfrac{1}{1 + \dfrac{1}{\sqrt{3}}} = \dfrac{1}{\dfrac{\sqrt{3}+1}{\sqrt{3}}} = \dfrac{\sqrt{3}}{\sqrt{3}+1}$

$\quad = \dfrac{\sqrt{3}(\sqrt{3}-1)}{(\sqrt{3}+1)(\sqrt{3}-1)} = \dfrac{3-\sqrt{3}}{3-1} = \dfrac{3-\sqrt{3}}{2}$

13. Counting strategies
1 WX, WY, WZ, XY, XZ, YZ
2 (A,B) (A,C) (A,D) (B,C) (B,D) (C,D) 6 games
3 9
4 Yes, there are 1 757 600 different possible codes
5 (a) (i) 10 000   (ii) 28 561
   (b) There are 900 possible codes

14. Problem-solving practice 1
1 60
2 (a) 3 packs of bread rolls, 5 packs of sausages, 8 packs of ketchup
   (b) 120
3 (a) 1000   (b) 720
   (c) $8^6 = 262\,144$ so 260 000 possible codes

15. Problem-solving practice 2
4 (a) 2.5×10^{25}
   (b) Underestimate, because you are dividing by a number that has been rounded up
5 3.1, because both values of UB and LB are the same for 2 s.f.
6 $6\sqrt{2} + 14$

ALBEBRA

16. Algebraic expressions
1 (a) m^3 (b) d^4 (c) e^5
2 (a) x^{11} (b) y^5 (c) t^4
3 (a) x^6 (b) y^{15} (c) t^{21}
4 (a) x^5 (b) y^9 (c) t^6
5 (a) $28x^3y^7$ (b) $2x^3y$ (c) $81x^8y^{20}z^{12}$
6 (a) $5x^3$ (b) $64x^{4.5}y^6$ (c) $3x^{1.25}y^{0.75}$
7 (a) $9x^8$ (b) $\dfrac{8x^2y^5}{5}$ (c) $\dfrac{16x^2y^6}{9}$

17. Expanding brackets
1 (a) $x^2 + 7x + 12$ (b) $x^2 + 2x - 15$ (c) $x^2 - 8x + 12$
2 (a) $x^2 + 6x + 9$ (b) $x^2 - 8x + 16$ (c) $4x^2 + 4x + 1$
3 (a) $x^3 + 8x^2 + 15x$ (b) $x^3 + 2x^2 - 8x$ (c) $x^3 - 10x^2 + 21x$
4 (a) $x^3 + 9x^2 + 27x + 27$
   (b) $x^3 - 12x^2 + 48x - 64$
   (c) $8x^3 + 12x^2 + 6x + 1$
5 (a) $6x^2 + 12x - 20$ (b) $x^3 + 3x^2 - 10x - 24$

18 Factorising
1 (a) $3(x + 2)$ (b) $2(p - 3)$ (c) $5(y - 3)$
2 (a) $x(x + 6)$ (b) $x(x + 4)$ (c) $x(x - 12)$
3 (a) $3p(p + 2)$ (b) $8y(y - 3)$
4 (a) $4d(d + 3)$ (b) $6x(x - 3)$
5 (a) $(x + 3)(x + 1)$ (b) $(x + 10)(x + 1)$
6 (a) $(x + 7)(x - 1)$ (b) $(x + 5)(x - 1)$ (c) $(x - 5)(x + 3)$
7 (a) $(x - 3)(x + 3)$ (b) $(x - 12)(x + 12)$
8 (a) $(3x - 1)(x - 2)$ (b) $(2x - 3)(x + 1)$ (c) $(3x + 2)(x - 6)$

19. Linear equations 1
1 (a) 4   (b) 6   (c) 5
   (d) 4   (e) 60   (f) -18
2 (a) 3   (b) 2   (c) -2
   (d) $-\frac{3}{5}$ (e) 1   (f) $-\frac{5}{4}$
3 (a) 4   (b) -7 (c) 12
   (d) 4   (e) 13   (f) 2
4 12

20. Linear equations 2
1 (a) $\frac{14}{3}$ (b) 4   (c) 3
2 (a) 1   (b) 12
3 (a) $\frac{55}{4}$ (b) 3
4 34

21. Formulae
1 (a) 8   (b) -19
2 (a) 23   (b) $\frac{35}{2}$
3 (a) 3500   (b) -200 (c) 8.84   (d) 19.8
4 26 000
5 $A = 2\pi r^2 + \dfrac{2V}{r}$

22. Arithmetic sequences
1 (a) $4n + 1$ (b) $3n - 1$ (c) $7n - 5$
2 $3n + 1$
3 $5n + 1$
4 (a) $4n - 1$ (b) Yes, because $n = 50$
5 (a) $6n - 3$
   (b) No, because n is not a whole number

23. Solving sequence problems
1 (a) $a = 3$ and $b = -1$
2 (a) 9   (b) $U_{n+1} = 2U_n + 3$
3 39
4 (a) 55   (b) $2x + 3y$ (c) $x = 5$ and $y = 6$
5 $27\sqrt{3}$, 81

24. Quadratic sequences
1 2, 7, 14, 23, 34, 47
2 (a) $n^2 + 2$ (b) $2n^2 + 4n - 3$
   (c) $2n^2 + 2$ (d) $3n^2 - 2n + 1$
   (e) $2n^2 + 4n - 4$ (f) $3n^2 - 5n + 3$
3 (a) $2n^2$
   (b) No; if $2n^2 = 75$ then n is not a whole number

25. Straight-line graphs 1
1 y values: 7, 6, 5, 4, 3, 2, 1, 0

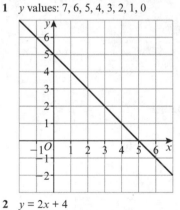

2 $y = 2x + 4$
3 $y = -2.5x + 20$

26. Straight-line graphs 2
1 (a) $y = 3x - 1$
   (b) $y = -2x + 12$
   (c) $y = 4x + 15$
2 (a) $y = 2x - 4$
   (b) $y = 2x + 1$
   (c) $y = 2x$
3 $y = 4x + 5$
4 $k = 5$

27. Parallel and perpendicular
1 (a) R   (b) S
2 (a) $y = 3x - 7$ (b) $y = 4 - \frac{1}{3}x$
3 $y = -\frac{1}{3}x + 8$
4 (a) $y = 2x + 6$ (b) $y = -\frac{1}{2}x + 6$
5 $y = -2x$

28. Quadratic graphs

1 (a) y values: 7, 2, −1, −2, −1, 2, 7

   (b)
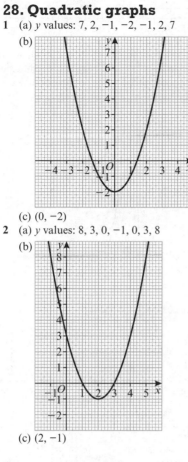

   (c) (0, −2)

2 (a) y values: 8, 3, 0, −1, 0, 3, 8

   (b)

   (c) (2, −1)

29. Cubic and reciprocal graphs

1 (a) y values: −2, 1, −2, −5, −2, 13

   (b)
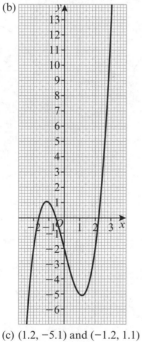

   (c) (1.2, −5.1) and (−1.2, 1.1)

   (d) −2.2, 0.3 and 1.9

2 (a) B   (b) D   (c) A   (d) C   (e) E

30. Real-life graphs

1 (a) 96   (b) Marseille

2
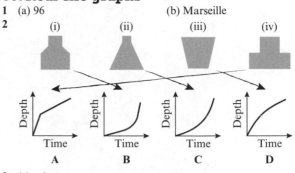

| A   B   C   D |

3 (a) −4

   (b) Depth of water decreases by 4 cm every hour

31. Quadratic equations

1 (a) 0 or 3   (b) 0 or $-\frac{5}{2}$

2 (a) 2 or −2   (b) $\frac{7}{4}$ or $-\frac{7}{4}$ (c) 5 or −5

3 (a) −2 or −4   (b) 3 or 4   (c) 3 or 7

4 (a) $-\frac{1}{2}$ or −3   (b) −1 or $-\frac{2}{3}$

5 (a) $(x + 5)(x − 2) = 60$
      $x^2 + 3x − 70 = 0$

   (b) length 12 cm, width 5 cm

6 1 or 2.5

32. The quadratic formula

1 (a) 1   (b) 2   (c) 0

2 (a) 4.56 or 0.438   (b) 0.281 or −1.78

   (c) 1.15 or −0.348   (d) −3.56 or 0.562

3 (a) $x(2x + 1) + 5x = 95$
      $2x^2 + 6x − 95 = 0$

   (b) 5.55

4 (a) $2x(x − 2) = 51$
      $2x^2 − 4x − 51 = 0$

   (b) 6.15

   (c) It has to be greater than 2 and cannot be negative

33. Completing the square

1 (a) $(x − 5)^2 − 24$ (b) $(x − 1)^2 + 6$ (c) $(x + 3)^2 − 13$

2 (a) $2(x + 3)^2 − 9$ (b) $3(x + 1)^2 − 5$ (c) $4(x + 1)^2 − 15$

3 (a) 1.16 and −5.16   (b) 6.54 and 0.459   (c) 0.25 and −2

4 (a) $-2 \pm \sqrt{7}$ (b) $\dfrac{-5 \pm \sqrt{73}}{4}$

5 (a) $3\left(x + \dfrac{5}{6}\right)^2 − \dfrac{13}{12}$ (b) $\dfrac{-5 \pm \sqrt{13}}{6}$

34. Simultaneous equations 1

1 (a) (3, 2)   (b) (5, −2)

2 (a)
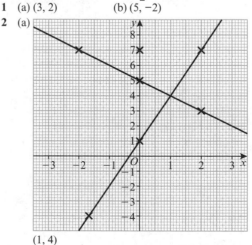

   (1, 4)

(b)

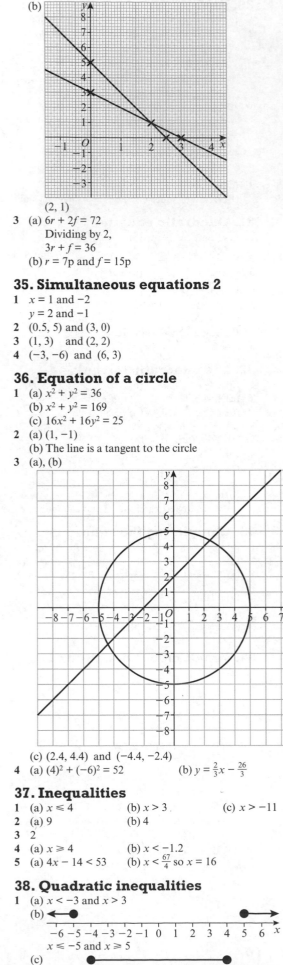

(2, 1)

3 (a) $6r + 2f = 72$
    Dividing by 2,
    $3r + f = 36$
  (b) $r = 7$p and $f = 15$p

35. Simultaneous equations 2

1 $x = 1$ and -2
  $y = 2$ and -1
2 (0.5, 5) and (3, 0)
3 (1, 3)   and (2, 2)
4 (−3, −6) and (6, 3)

36. Equation of a circle

1 (a) $x^2 + y^2 = 36$
  (b) $x^2 + y^2 = 169$
  (c) $16x^2 + 16y^2 = 25$
2 (a) (1, −1)
  (b) The line is a tangent to the circle
3 (a), (b)

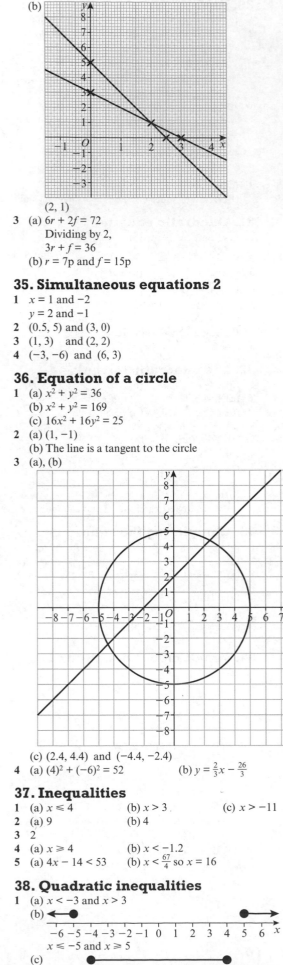

(c) (2.4, 4.4) and (−4.4, −2.4)
4 (a) $(4)^2 + (-6)^2 = 52$ (b) $y = \frac{2}{3}x - \frac{26}{3}$

37. Inequalities

1 (a) $x \leqslant 4$ (b) $x > 3$ (c) $x > -11$
2 (a) 9   (b) 4
3 2
4 (a) $x \geqslant 4$ (b) $x < -1.2$
5 (a) $4x - 14 < 53$ (b) $x < \frac{67}{4}$ so $x = 16$

38. Quadratic inequalities

1 (a) $x < -3$ and $x > 3$
  (b)

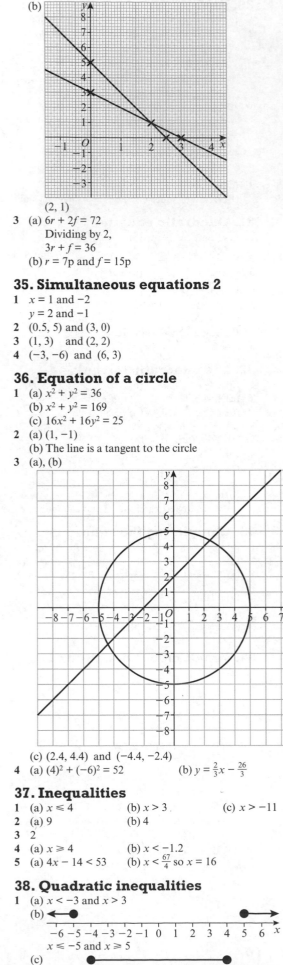

$x \leqslant -5$ and $x \geqslant 5$

  (c)

$x \geqslant -4$ and $x \leqslant 4$

(d)

$x < -1$ and $x > 1$
2 (a) $x < 3\sqrt{3}$ and $x > -3\sqrt{3}$ (b) $x \leqslant -4\sqrt{3}$ and $x \geqslant 4\sqrt{3}$
3 (a) $x > -1$ and $x < 5$ (b) $x \leqslant 3$ and $x \geqslant 8$
  (c) $x \geqslant -2$ and $x \leqslant 5$ (d) $x < \frac{1}{2}$ and $x > \frac{3}{2}$
4 (a)

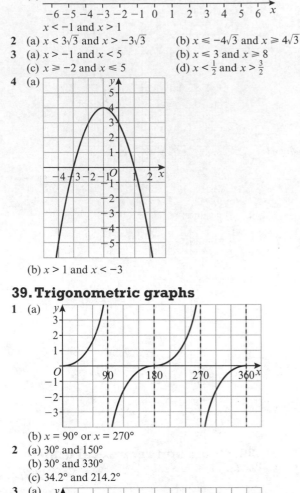

(b) $x > 1$ and $x < -3$

39. Trigonometric graphs

1 (a)

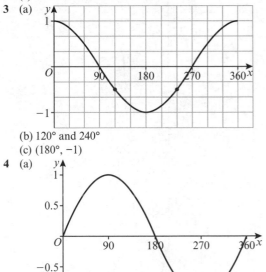

(b) $x = 90°$ or $x = 270°$
2 (a) 30° and 150°
  (b) 30° and 330°
  (c) 34.2° and 214.2°
3 (a)

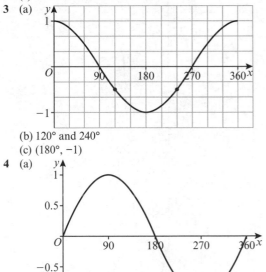

(b) 120° and 240°
(c) (180°, −1)
4 (a)

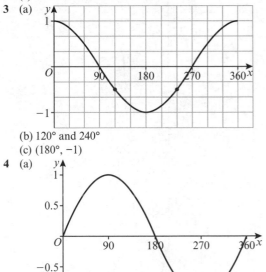

(b) 60° and 120°
(c) (90°, 1) and (270°, −1)

40. Transforming graphs

1 (a)

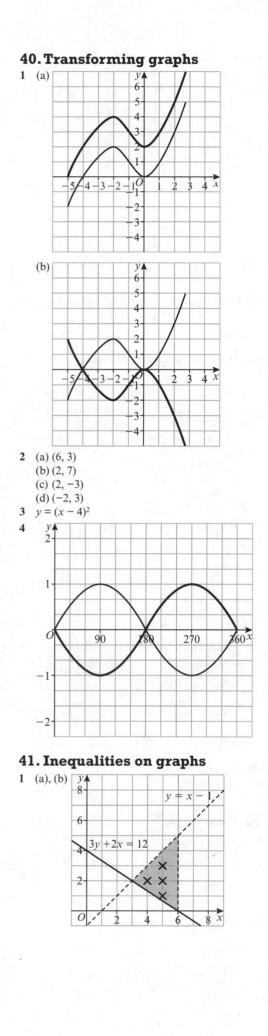

(b)

2 (a) (6, 3)
(b) (2, 7)
(c) (2, −3)
(d) (−2, 3)

3 $y = (x − 4)^2$

4

41. Inequalities on graphs

1 (a), (b)

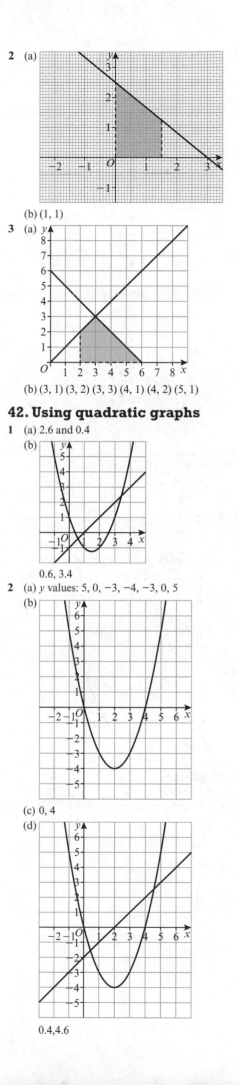

2 (a)

(b) (1, 1)

3 (a)

(b) (3, 1) (3, 2) (3, 3) (4, 1) (4, 2) (5, 1)

42. Using quadratic graphs

1 (a) 2.6 and 0.4

(b)

0.6, 3.4

2 (a) y values: 5, 0, −3, −4, −3, 0, 5

(b)

(c) 0, 4

(d)

0.4, 4.6

43. Turning points

1 (a) $(-2, -10)$ (b) $(-1, -5)$
  (c) $(-\frac{3}{2}, -\frac{1}{4})$ (d) $(-\frac{3}{4}, -\frac{65}{8})$

2 (a) $p = -2$ and $q = -2$ (b) -2

3 (a)

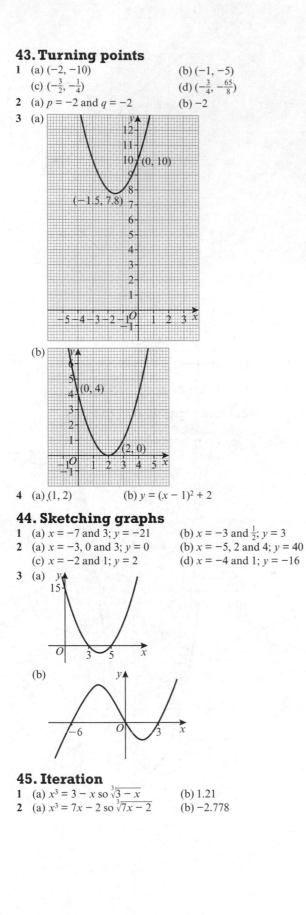

  (b)

4 (a) $(1, 2)$ (b) $y = (x - 1)^2 + 2$

44. Sketching graphs

1 (a) $x = -7$ and 3; $y = -21$ (b) $x = -3$ and $\frac{1}{2}$; $y = 3$
2 (a) $x = -3, 0$ and 3; $y = 0$ (b) $x = -5, 2$ and 4; $y = 40$
  (c) $x = -2$ and 1; $y = 2$ (d) $x = -4$ and 1; $y = -16$

3 (a)

  (b)

45. Iteration

1 (a) $x^3 = 3 - x$ so $\sqrt[3]{3 - x}$ (b) 1.21
2 (a) $x^3 = 7x - 2$ so $\sqrt[3]{7x - 2}$ (b) -2.778

3 (a) y values: $-20, -6, -4, -8, -12, -10, 4$
  (b)

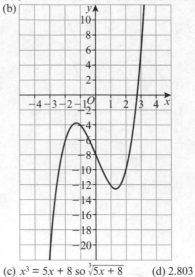

  (c) $x^3 = 5x + 8$ so $\sqrt[3]{5x + 8}$ (d) 2.803

46. Rearranging formulae

1   $x = 3(y - 4)$

2   $h = \dfrac{2d^2}{3}$

3   $w = \dfrac{120P - 10}{3}$

4   $x = \dfrac{P - \pi r - 2r}{2}$

5   $x = \dfrac{Mt - t^2}{1 - M}$

6   $t = \dfrac{yp}{2p + y}$

47. Algebraic fractions

1 (a) $\dfrac{7}{8x}$ (b) $\dfrac{x + 7}{6}$ (c) $\dfrac{x - 7}{(x + 3)(x - 2)}$ (d) $\dfrac{x^2 + y^2}{xy}$

2 (a) $\dfrac{x + 1}{4}$ (b) $\dfrac{x + 5}{x + 3}$

3 (a) $\dfrac{3}{5}$ (b) $\dfrac{8}{5r}$ (c) $\dfrac{5}{x^2}$ (d) $\dfrac{x - 4}{3}$

4 (a) $x = \dfrac{8}{3}$ (b) $x = \dfrac{1}{16}$ (c) $x = \dfrac{18}{13}$ (d) $x = 6$

48. Quadratics and fractions

1 (a) $x = 2$ and $x = 6$ (b) $x = 3.5$ and $x = 5$
  (c) $x = 0$ and $x = -2$ (d) $x = \frac{3}{8}$ and $x = 1$
  (e) $x = \frac{1}{3}$ and $x = \frac{1}{2}$ (f) $x = 13$ and $x = -5$
  (g) $x = -1$ and $x = 4$

2 (a) $\dfrac{400}{x}$ (b) $\dfrac{400}{x + 10}$

  (c) $\dfrac{400}{x} - \dfrac{400}{x + 10} = \dfrac{40}{60}$ (d) 72.6 km/h

49. Surds 2

4   $a = 7$ and $b = 8$
5 (a) $\dfrac{1}{2}$ (b) $3\sqrt{2}$

50. Functions

1 (a) (i) 13 (ii) $\frac{1}{3}$ (b) $\dfrac{2}{2x - 1}$

2 (a) -11 (b) 3 (c) 126 (d) $4x^3 + 1$

3 (a) 3 (b) $\dfrac{x}{4 - x}$

4   $x = 1$ and $x = \frac{1}{2}$

51. Inverse functions

1 (a) $\dfrac{x + 5}{3}$ (b) $\dfrac{3(x - 7)}{4}$ (c) $\dfrac{7}{1 - x}$

2 (a) $\dfrac{x - 4}{3}$ (b) $\dfrac{x + 1}{4}$ (c) $\dfrac{7x - 13}{12}$

3 (a) $\dfrac{x - 1}{2}$ (b) 5

4 (a) $\dfrac{5}{3}$ (b) $\dfrac{x}{x - 1}$

52. Algebraic proof

1 LHS $= 4n^2 - 2n - 2n + 1 + 4n^2 + 2n + 2n + 1$
   $= 8n^2 + 2 = 2(4n^2 + 1) = $ RHS

2 $5x - 5c = 4x - 5$
   $x = 5c - 5$
   $x = 5(c - 1)$

3 $(3x + 1)^2 - (3x - 1)^2$
   $= (3x + 1)(3x + 1) - (3x - 1)(3x - 1)$
   $= 9x^2 + 3x + 3x + 1 - (9x^2 - 3x - 3x + 1)$
   $= 9x^2 + 3x + 3x + 1 - 9x^2 + 3x + 3x - 1$
   $= 12x = 4(3x)$

4 $2n + 2n + 2 + 2n + 4 = 6n + 6 = 6(n + 1)$

5 $n(n + 1) + (n + 1)(n + 2) = n^2 + n + n^2 + 2n + n + 2$
   $= 2n^2 + 4n + 2 = 2(n^2 + 2n + 1)$
   $= 2(n + 1)^2$

6 $\dfrac{1}{n} - \dfrac{1}{n + 1} = \dfrac{n + 1 - n}{n(n + 1)} = \dfrac{1}{n(n + 1)}$

7 $(10a + b)^2 - (10b + a)^2$
   $= 100a^2 + 20ab + b^2 - 100b^2 - 20ab - a^2$
   $= 99a^2 - 99b^2 = 99(a^2 - b^2)$

53. Exponential graphs

1 (a) (b) (c)

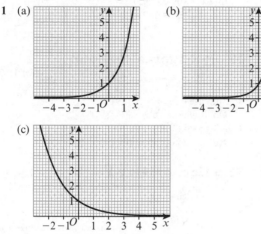

2 $a = 4$ and $k = 1$

3 (a) $2000 \times 1.2^3 = 3456$
   (b) $a = 2000$ and $b = 1.2$
   (c) $k = 1.728$

54. Gradients of curves

1 (a) 1.1

2 (a)

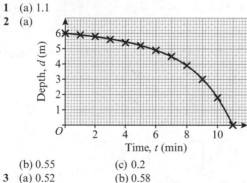

   (b) 0.55   (c) 0.2

3 (a) 0.52   (b) 0.58

55. Velocity–time graphs

1 (a) 1.5 m/s^2 (b) 0.75 m/s^2 (c) 375 m

2 (a)

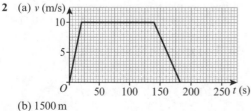

   (b) 1500 m

3 (a) 3.2
   (b) It is the acceleration of the object at $t = 3$ seconds

56. Areas under curves

1 (a) 19.5
   (b) Underestimate, because the trapeziums are below the curve

2 (a) 6.75
   (b) Underestimate, because the trapeziums are below the curve

3 28.5

57. Problem-solving practice 1

1 $x = 4$

2 $x = 5$ and $y = 14$

3 D is at $(3, 6)$
   $m_{AB} = \frac{3}{2}$, $m_{CD} = -\frac{2}{3}$, $m_{AB} \times m_{CD} = -1$
   So, no she is incorrect, CD is perpendicular to AB

4 $n^2 - 3n + 3$

58. Problem-solving practice 2

5 1 m

6 Area under graph is 47 m
   No, the driver does not stop before the traffic lights

7 $(4, 7)$ and $(-1, -8)$

RATIO & PROPORTION

59. Calculator skills 2

1 £86.40

2 (a) 31 200 000   (b) 13.9%

3 (a) 12.9%   (b) 19 000 000

4 (a) 675 euros   (b) £96
   (c) Yes, because the calculations can be done without a calculator and 1.5 euros to the pound is a good approximation.

60. Ratio

1 (a) £20 : £30   (b) £100 : £250 : £400

2 (a) 5 parts = £120
       1 part = £24
       Amish = $3 \times £24 = £72$
   (b) $(3 \times 24) + (4 \times 24) + (9 \times 24) = £384$

3 28 g

4 £4400

61. Proportion

1 20 days

2 Large basket

3 8

4 No, because she has only 200 ml of milk and needs 250 ml

5 Germany: 4.95 lb = £20.37 so 1 lb = £4.11
   Scotland: 2.75 lb = £12.42 so 1 lb = £4.52
   Cheaper in Germany

62. Percentage change

1 Mega-jet is cheaper

2 (a) 34.8%   (b) 28%

3 £23 320

4 £1454.64

5 33.3%

63. Reverse percentages

1 £60

2 £33 250

3 £33 600

4 750

5 No, she earned £518.52 last year

6 Alison invested £1650 and Nav invested £1680, so Nav invested more

64. Growth and decay

1 £17 569.20

2 3

3 (a) £12 528.15   (b) £6332.78

4 (a) 6.5%   (b) £2815.71

5 Terry is wrong. It is worth £110.74 after 4 years.

65. Speed

1. 425 km/h
2. 8.9 m/s
3. 96 km/h
4. 240 km
5. Pavan's speed is 35 ÷ 0.25 = 140 km/h
6. Jane's speed is 40 km/h and Carol's speed is 29.1 km/h, so Carol has the lower average speed
7. 100 m speed is 6.75 m/s and 200 m speed is 6.72 m/s, so 100 m race had the faster speed

66. Density

1. 0.875 g/cm^3
2. 147 g
3. 10 375 cm^3
4. 432 g
5. 5666.4 g
6. 950 ÷ 96 = 9.9, so Gavin is not correct; it is bronze
7. 7.374 g/cm^3

67. Other compound measures

1. 400 N/m^2
2. 18 750 N/m^2
3. 0.006 25 m^2
4. Volume = 120 000 ÷ 2 = 60 000
   Time = 60 000 ÷ 250 = 240 s = 4 min
5. No, population is 12 170 × 0.24 = 2921
6. It takes 13 hours and 20 min; she gets there at 10.20pm; yes, can reach the peak by 11pm

68. Proportion and graphs

1. 1170
2. 7.5
3. (a) 0.16   (b) 5
   (c) The graph is a straight line through the origin; there is a constant increase / as extension increases force increases
4. (a) 2   (b) 50
   (c) As pressure increases volume decreases

69. Proportionality formulae

1. (a) 1050 s   (b) 600 s
2. 14.4 g
3. (a) $T = 22x$ (b) 308 N   (c) 14.5 cm
4. (a) 84   (b) 245

70. Harder relationships

1. (a) $d = 4t^2$ (b) 144   (c) 4
2. (a) $S = \dfrac{3240}{f^2}$ (b) 50.625
3. (a) $I = 0.75\sqrt{P}$ (b) 5.25   (c) 44.4
4. 2

71. Problem-solving practice 1

1. Nile: £32.57
   T-bay: £37.08
   Nile is cheaper
2. (a) 8 parts = £240
       1 part = £30
       Angus = 4 × 30 = £120
   (b) (4 × 30) + (7 × 30) + (12 × 30) = £690
3. 30 ml

72. Problem-solving practice 2

4. 325 ÷ $3\frac{7}{15}$ = 93.75 km/h
   Yes, he breaks the speed limit
5. 1000 g
6. Yes, he saves £17 674.68
7. Yes, she is correct

GEOMETRY & MEASURES

73. Angle properties

1. $x = 18$, because angles on a straight line add up to 180, and $y = 68$, because alternate angles are equal.

2. $x = 14$ and $y = 8$
3. 17°

74. Solving angle problems

1. $x = 24$ and $y = 24$
2. (a) 20°   (b) angle DAC = angle DCA
3. $x = 36°$
   Base angles in an isosceles triangle are equal
   Angle ADB = angle DBC are alternate angles
4. (a) $2x$
   (b) Angles in a triangle add up to 180°
   Base angles in an isosceles triangle are equal
   Angle $CDB = (180 - 2x) \div 2 = 90 - x$
   and angle $BDA = x$
   Angle $CDA = 90 - x + x = 90$

75. Angles in polygons

1. (a) 40°   (b) 9
2. (a) 12   (b) 10
3. 135
4. (a) 18   (b) 2880°
5. Exterior angle = 60°
   Interior angle = 120°
   $x + 120 + x = 180$
   $x = 30$

76. Pythagoras' theorem

1. (a) 10.7   (b) 6.75   (c) 15.7
2. 13.47
3. Suitcase diagonal = 119 cm
   No, it will not fit
4. Diagonal of cross-section of pool = 5.8 m
   No, it cannot be totally immersed

77. Trigonometry 1

1. (a) 54.3   (b) 57.8   (c) 31.1
2. 62.8°
3. 53.0
4. She can use smooth tiles on her roof, because x is 20.4 and hence the angle is greater than 17°
5. $y = 31.8$, so no he cannot

78. Trigonometry 2

1. (a) 14.3   (b) 16.3   (c) 19.1
2. 5.35 m
3. 55.1 m
4. (a) 21.0 m   (b) 60.3°

79. Solving trigonometry problems

1.

	0°	30°	45°	60°	90°
sin	0	$\frac{1}{2}$	$\frac{1}{\sqrt{2}}$	$\frac{\sqrt{3}}{2}$	1
cos	1	$\frac{\sqrt{3}}{2}$	$\frac{1}{\sqrt{2}}$	$\frac{1}{2}$	0
tan	0	$\frac{1}{\sqrt{3}}$	1	$\sqrt{3}$	–

2. (a) 9   (b) 4.9   (c) 16.9
3. (a) 30°   (b) 30°   (c) 30°
4. $30\sqrt{3}$
5. Height above ground is opposite angle of elevation and length of plank is hypotenuse.
   Angle of elevation = $\sin^{-1}(3 \div 6)$ = $\sin^{-1}(0.5)$ = 30°

80. Perimeter and area

1. (a) 53 cm   (b) 148 cm^2
2. 44 cm
3. (a) $2(3x - 2) + (3x - 2)(2x + 5) = 25$
   $6x^2 + 17x - 39 = 0$
   (b) $x = 1.5$ and $-13/3$ (c) 25
4. 250 cm^2

81. Units of area and volume

1 (a) $60\,000\,\text{cm}^2$ (b) $1500\,\text{mm}^2$ (c) $4\,000\,000\,\text{m}^2$
   (d) $50\,\text{m}^2$ (e) $600\,\text{cm}^2$ (f) $0.8\,\text{km}^2$

2 (a) $22\,000\,000\,\text{cm}^3$ (b) $28\,000\,\text{mm}^3$
   (c) $3\,000\,000\,000\,\text{m}^3$ (d) $200\,\text{m}^3$
   (e) $50\,000\,\text{cm}^3$ (f) $0.42\,\text{km}^3$

3 (a) 200 litres   (b) 8000 litres   (c) 12 000 litres

4 (a) 30 litres   (b) 96 000 litres

5 7200 N

82. Prisms

1 (a) $48\,\text{cm}^3$ (b) $144\,\text{mm}^3$ (c) $384\,\text{mm}^3$

2 (a) $108\,\text{cm}^2$ (b) $216\,\text{mm}^2$ (c) $339.2\,\text{mm}^2$

3 $352\,\text{cm}^3$

4 $x = 10\,\text{cm}$

83. Circles and cylinders

1 3140 cm

2 40.02 = 40

3 Area $A = \frac{1}{4} \times \pi \times x^2 = \frac{1}{4}\pi x^2$

   Area $B = 2 \times \frac{1}{2} \times \pi \times \left(\frac{x}{2}\right)^2 = \frac{1}{4}\pi x^2$

4 Cylinder volume $= \pi \times 15^2 \times 18 = 12\,723\,\text{cm}^3$
   Cube volume $= 24 \times 24 \times 24 = 13\,824\,\text{cm}^3$
   Volume of cube is greater

5 Area of A $= (2 \times \pi \times 9 \times 21) + (\pi \times 9^2) = 1442\,\text{cm}^2$
   Area of B $= (8 \times 8) + (4 \times 8 \times 42) = 1408\,\text{cm}^2$
   Vase A has the greatest surface area

84. Sectors of circles

1 (a) 5.59 cm   (b) 27.2 mm

2 (a) 24.4 cm   (b) 50.2 mm

3 (a) $36.3\,\text{cm}^2$ (b) $145\,\text{mm}^2$

4 $66.4\,\text{cm}^2$

5 $75\,\text{cm}^2$

85. Volumes of 3D shapes

1 (a) $251\,\text{cm}^3$ (b) $7240\,\text{cm}^3$ (c) $142\,\text{cm}^3$

2 (a) $302\,\text{cm}^3$ (b) $300\,\text{cm}^3$

3 $\pi \times x^2 \times 3x = \frac{1}{3} \times \pi \times x^2 \times h$
   So $h = 9x$

4 $\pi \times x^2 \times h = 3 \times \frac{4}{3} \times \pi \times x^3$
   So $h = 4x$

86. Surface area

1 (a) $251\,\text{cm}^2$ (b) $1810\,\text{cm}^2$ (c) $462\,\text{cm}^2$

2 (a) $226\,\text{cm}^2$ (b) $650\,\text{cm}^2$ (c) $320\,\text{cm}^2$

3 $427\,\text{cm}^2$

87. Plans and elevations

1 (b)

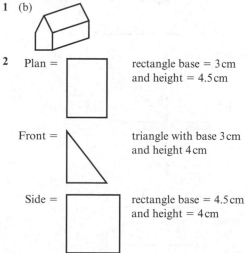

2 Plan =   rectangle base = 3 cm
            and height = 4.5 cm

   Front =   triangle with base 3 cm
            and height 4 cm

   Side =   rectangle base = 4.5 cm
            and height = 4 cm

3

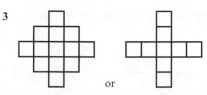

or

88. Translations, reflections and rotations

1 (b), (c)

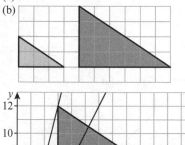

2 (a) Reflection in the line $x = -1$
   (b) Rotation 90° clockwise about (2, 0)
   (c) Translation $\begin{bmatrix} 6 \\ -6 \end{bmatrix}$

89. Enlargement

1 (a) 5

   (b)

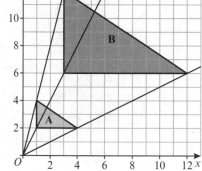

2

3 (a), (b)

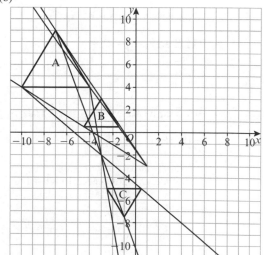

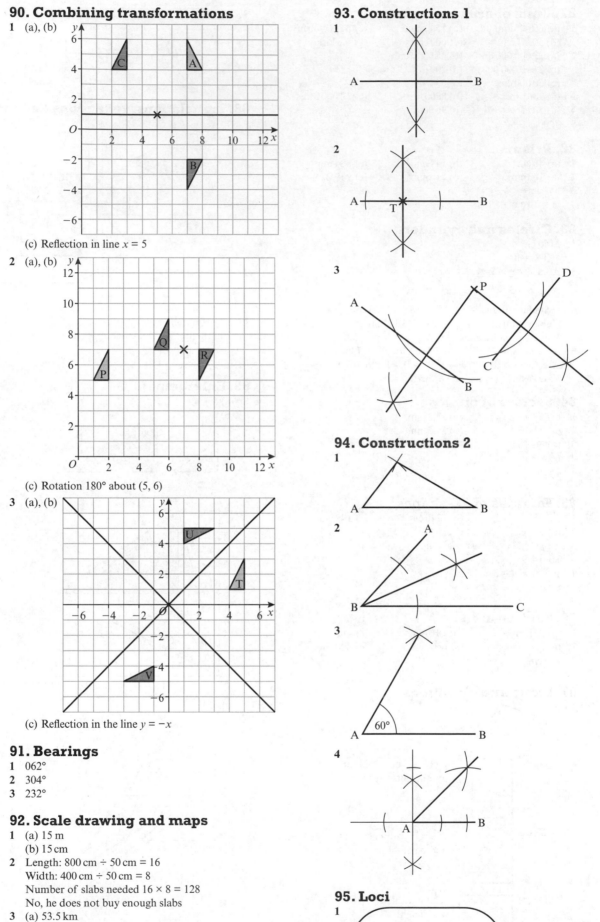

90. Combining transformations

1 (a), (b)

(c) Reflection in line $x = 5$

2 (a), (b)

(c) Rotation 180° about (5, 6)

3 (a), (b)

(c) Reflection in the line $y = -x$

91. Bearings

1 062°
2 304°
3 232°

92. Scale drawing and maps

1 (a) 15 m
  (b) 15 cm
2 Length: 800 cm ÷ 50 cm = 16
  Width: 400 cm ÷ 50 cm = 8
  Number of slabs needed 16 × 8 = 128
  No, he does not buy enough slabs
3 (a) 53.5 km
  (b) 8 km

93. Constructions 1

1

2

3

94. Constructions 2

1

2

3

4

95. Loci

1

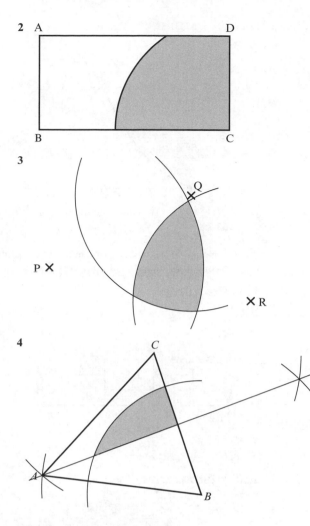

2

3

4

96. Congruent triangles
1 alternate; parallelogram; SAS
2 BC is common; $BM = CL$; angle LBC = angle MCB; SAS
3 (a) $AB = AC$ (equilateral triangle); AD is common; angle ADC = angle ADB (90°); triangle ADC is congruent to triangle ADB; RHS
   (b) Since triangles ADB and ADC are congruent, $BD = DC$ so DC is half of BC. But, $BC = AC$ (equilateral triangle) so DC is half of AC.
4 Angle EDC = angle EBA (alternate); angle BAE = angle DCE (alternate); $DE = EB$; ABE and CDE are congruent; AAS

97. Similar shapes 1
1 (a) 30   (b) 18
2 35
3 12.5
4 5.6

98. Similar shapes 2
1 135
2 (a) 2430   (b) 18
3 5375
4 (a) 180 cm   (b) 268 cm³

99. The sine rule
1 (a) 4.90 cm   (b) 5.37 cm
2 (a) 29.3°   (b) 34.3°
3 22.2 m
4 26.5 m

100. The cosine rule
1 (a) 9.60 cm   (b) 9.85 cm
2 (a) 51.9°   (b) 141.6°
3 16.0 cm
4 60.3 km

101. Triangles and segments
1 (a) 71.9 cm²   (b) 72.1 cm²
2 (a) 6.23 cm²   (b) 229 cm²
3 (a) 43.3 m²   (b) 30.2 m²

102. Pythagoras in 3D
1 13 cm
2 13.3 cm
3 7.42 cm
4 Length of diagonal is 32.9 cm, so stick will not fit

103. Trigonometry in 3D
1 (a) 13 cm   (b) 24.8°
2 58.0°
3 15.0 cm

104. Circle facts
1 80°
2 (a) (i) 90°   (ii) Tangent to circle is 90°
   (b) 42°
3 Angle $OBA = 90 - x$
   Angle $OAB = 90 - x$
   Angle $AOB = 180° - (90 - x + 90 - x) = 2x$

105. Circle theorems
1 63° because angles in the same segment are equal
2 (a) 148° because the angle at the centre is twice the angle at the circumference
   (b) 106° because opposite angles in a cyclic quadrilateral add up to 180°
3 (a) 54° because tangent to circle is at 90°
   (b) 54° because angle in a semicircle is 90°
4 Angle $ABC = 54°$ (alternate segment theorem)
   Angle $BCA = 63°$ (base angles in an isosceles triangle are equal)
   Angle $OCA = 36°$ (tangent to circle is at 90°)
   Angle $OCB = 27°$

106. Vectors
1 (a) $\begin{pmatrix} 2 \\ 5 \end{pmatrix}$ (b) $\begin{pmatrix} -2 \\ -5 \end{pmatrix}$ (c) $\begin{pmatrix} 5 \\ 3 \end{pmatrix}$
   (d) $\begin{pmatrix} -5 \\ -3 \end{pmatrix}$ (e) $\begin{pmatrix} 5 \\ -6 \end{pmatrix}$ (f) $\begin{pmatrix} -5 \\ 6 \end{pmatrix}$
2 (a) $\mathbf{a} + \mathbf{b}$ (b) $-\mathbf{a} - \mathbf{b}$
3 (a) $\mathbf{q} + \mathbf{p}$ (b) $-\mathbf{q} - \mathbf{p}$ (c) $\mathbf{p} - \mathbf{q}$ (d) $\mathbf{q} - \mathbf{p}$
4 (a) $\mathbf{a} + \mathbf{b}$ (b) $-\mathbf{a} - \mathbf{b}$

107. Vector proof
1 (a) (i) $2\mathbf{b}$ (ii) $2\mathbf{b} - \mathbf{a}$
   (b) $\overrightarrow{CA} = 2\mathbf{a} - 4\mathbf{b} = -2(2\mathbf{b} - \mathbf{a})$, this is a multiple of $(2\mathbf{b} - \mathbf{a})$ so CA is parallel to MN
2 (a) (i) $\mathbf{b} - \mathbf{a}$ (ii) $\frac{1}{5}(\mathbf{b} - \mathbf{a})$ (iii) $\frac{1}{5}(4\mathbf{a} - \mathbf{b})$
   (b) $\frac{1}{2}\mathbf{b} - \frac{4}{5}(\mathbf{b} - \mathbf{a}) = \frac{1}{10}(8\mathbf{a} - 3\mathbf{b})$
3 $\overrightarrow{AD} = \mathbf{c} - \mathbf{a}$
   $\overrightarrow{OE} = 2\mathbf{c} - 2\mathbf{a}$
   $\overrightarrow{OE} = 2(\mathbf{c} - \mathbf{a}) = 2\overrightarrow{AD}$
   Ratio of lengths $AD : OE$ is $1 : 2$

108. Problem-solving practice 1
1 $\frac{1}{3} \times \pi \times x^2 \times h = \frac{4}{3} \times \pi \times x^3$
   So $h = 4x$
2 (a) 37.5 km   (b) 271°
3 $450 \div \left(\frac{3}{5}\right)^2 = 162$ cm²

109. Problem-solving practice 2
4 55.9°
5 $\overrightarrow{BC} = 4\mathbf{a} + 8\mathbf{b}$
   $\overrightarrow{DB} = 12\mathbf{a} - 8\mathbf{b}$
   $\overrightarrow{AE} = 3\mathbf{a} + 6\mathbf{b}$
   $\overrightarrow{BC}$ and $\overrightarrow{AE}$ are multiples of $\mathbf{a} + 2\mathbf{b}$, so lines BC and AE are parallel
6 29.6 cm²

PROBABILITY & STATISTICS

110. Mean, median and mode
1 67
2 7, 7, 8, 11, 12
3 $X = 2$ and $Y = 16$
4 32

111. Frequency table averages
1 (a) 1   (b) 1   (c) 1.6   (d) 4
2 (a) 5.6
   (b) Working with grouped data

112. Interquartile range
1 (a) 33   (b) 22
2 (a) 43   (b) 20
3 (a) 19   (b) 50   (c) 38   (d) 29

113. Line graphs
1 (a)

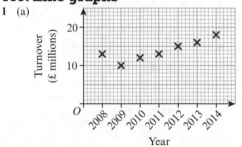

   (b) Upwards
2 (a) 4
   (b) 60
3 (a) 6
   (b) 6.2

114. Scatter graphs
1 (a) Positive
   (b) 135 g
   (c) 235 g
   (d) 80 pages is out of the data range
2 (a) Negative   (b) 1250
   (c) Reliable, as it is within the data range
   (d) Gives a negative price

115. Sampling
1 (a) It is quick, cheap and easier to handle
   (b) Not very reliable, because sample is small
   (c) Ask more people on different days
2 (a) 21.5
   (b) 1.7
   (c) Part (a), because it is in the data range
   (d) Carry out more experiments

116. Stratified sampling
1 14
2 132
3 14, 10, 21, 20
   (One of these needs to be rounded down, so an alternative
   answer is 14, 10, 22, 19)
4 5

117. Capture–recapture
1 525
2 750
3 500

118. Cumulative frequency
1 (a) 48 g
   (b) 27 g

2 (a)

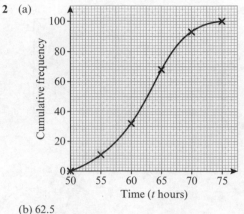

   (b) 62.5
   (c) 7.5
   (d) 68 hours = 86%, so 14% last longer, not 20%. Tom is
   incorrect

119. Box plots
1 (a) 44 cm
   (b) 12 cm
   (c) 21
2

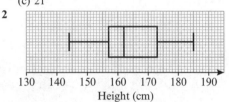

3 (a) 75%
   (b) 60

120. Histograms
1 (a) 15, 18
   (b)

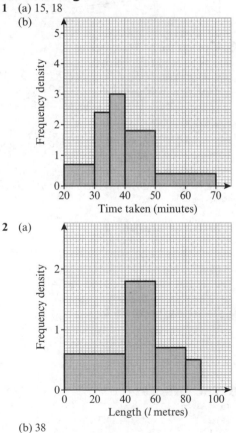

2 (a)

   (b) 38

121. Frequency polygons

1 (a)

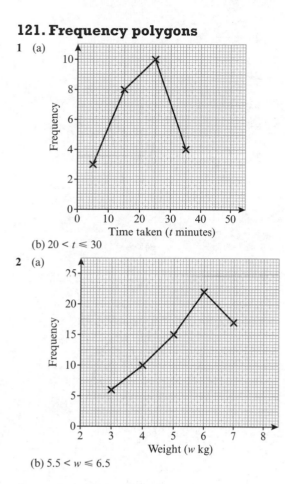

(b) $20 < t \leqslant 30$

2 (a)

(b) $5.5 < w \leqslant 6.5$

122. Comparing data

1 The median in class 11A is lower than the median in class 11B. The range for both classes is the same

2 Anjali has a larger range/IQR. Carol has a higher median

123. Probability

1 (a) 0.7   (b) 0.3
2 0.21
3 (a) 0.62   (b) 0.21
4 (a) 0.68   (b) 0.08

124. Relative frequency

1 (a) $\dfrac{53}{302}$ (b) $\dfrac{140}{302}$

2 (a) $\dfrac{1}{5}$ (b) $\dfrac{17}{50}$ (c) $\dfrac{41}{50}$

3 (a) $\dfrac{143}{202}$

(b) Quite accurate, because a large sample

125. Venn diagrams

1 (a) (i) 15   (ii) Maths
   (b) (i) 9   (ii) Do not study French nor German
2 (a) 0.6   (b) 0.5   (c) 0.4
3 (a) $\dfrac{1}{8}$ (b) $\dfrac{11}{40}$
4 (a)

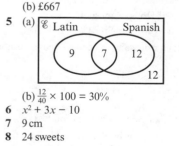

(b) $\dfrac{12}{30}$ (c) $\dfrac{14}{30}$

126. Conditional probability

1 (a) $\dfrac{85}{219}$ (b) $\dfrac{85}{183}$

2 (a) $\dfrac{26}{38}$ (b) $\dfrac{26}{46}$

3 (a) $\dfrac{9}{39}$ (b) $\dfrac{25}{62}$

127. Tree diagrams

1 (a)

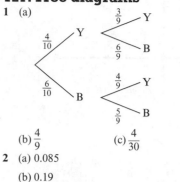

(b) $\dfrac{4}{9}$ (c) $\dfrac{4}{30}$ (d) $\dfrac{8}{15}$

2 (a) 0.085

  (b) 0.19

128. Problem-solving practice 1

1 (a) (i) 0.75   (ii) 0.2
  (b) 30
2 The median height in Park A is greater the median height in Park B. The range in Park B is greater than in Park A
3 (a) $\dfrac{7}{25}$ (b) $\dfrac{12}{25}$

129. Problem-solving practice 2

4 $\dfrac{27}{43}$

5 15

6 (a) $\dfrac{31}{80}$ (b) $\dfrac{17}{28}$

MATHS PRACTICE EXAM PAPER

Paper 1

1 (a) 9
  (b) 36
2

3	9
4	1 3 5 6 9
5	2 3 3 4 6 8 9
6	2 4 5

3 | 9 means 39 mph

3 (a) (i) 42°   (ii) Alternate angles
  (b) 69°
4 (a)

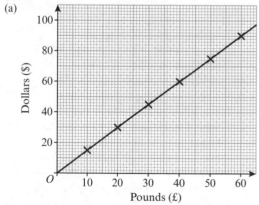

(b) £667
5 (a)

Latin   Spanish
9   7   12
12

(b) $\dfrac{12}{40} \times 100 = 30\%$
6 $x^2 + 3x - 10$
7 9 cm
8 24 sweets

9

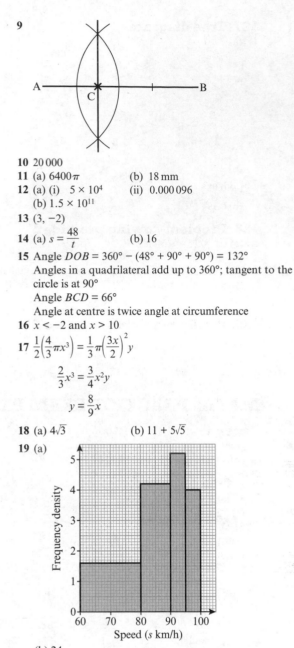

10 20 000

11 (a) 6400π (b) 18 mm

12 (a) (i) 5×10^4 (ii) 0.000 096

    (b) 1.5×10^{11}

13 $(3, -2)$

14 (a) $s = \dfrac{48}{t}$ (b) 16

15 Angle $DOB = 360° - (48° + 90° + 90°) = 132°$

    Angles in a quadrilateral add up to 360°; tangent to the circle is at 90°

    Angle $BCD = 66°$

    Angle at centre is twice angle at circumference

16 $x < -2$ and $x > 10$

17 $\dfrac{1}{2}\left(\dfrac{4}{3}\pi x^3\right) = \dfrac{1}{3}\pi\left(\dfrac{3x}{2}\right)^2 y$

        $\dfrac{2}{3}x^3 = \dfrac{3}{4}x^2 y$

          $y = \dfrac{8}{9}x$

18 (a) $4\sqrt{3}$ (b) $11 + 5\sqrt{5}$

19 (a)

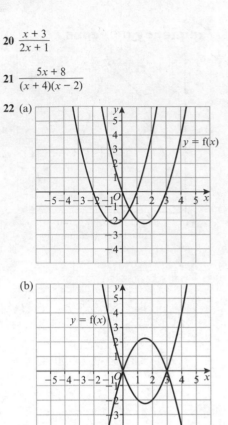

    (b) 24

20 $\dfrac{x + 3}{2x + 1}$

21 $\dfrac{5x + 8}{(x + 4)(x - 2)}$

22 (a)

(b)

Published by Pearson Education Limited, 80 Strand, London, WC2R 0RL.

www.pearsonschoolsandfecolleges.co.uk

Copies of official specifications for all Edexcel qualifications may be found on the website: www.edexcel.com

Text and illustrations © Pearson Education Limited 2016

Edited by Project One Publishing Solutions, Scotland

Typeset and illustrations by Tech-Set Ltd, Gateshead

Produced by Out of House Publishing

Cover illustration by Miriam Sturdee

The right of Navtej Marwaha to be identified as author of this work has been asserted by him in accordance with the Copyright, Designs and Patents Act 1988.

First published 2016

18 17 16
10 9 8 7 6 5 4 3

British Library Cataloguing in Publication Data

A catalogue record for this book is available from the British Library

ISBN 978 1 447 98793 2